THINKING MEDICINE

THINKING MEDICINE

Structure Your Knowledge
for Success in Medical Exams

Dr Cristina Koppel
Dr Andreas Naparus

www.thinkingmedicine.com

Thinking Medicine

Structure Your Knowledge for Success in Medical Exams

Published by Cavaye Publishing

Second Edition

Version 1

ISBN 978-0-9561288-1-2

The authors have taken care to ensure that the information given in this text is from reliable sources and up to date. However, while every measure has been made to ensure accuracy, it remains with the reader to confirm that the information complies with the latest standards of practice. No responsibility can be accepted by the authors or publishers with regards to the actions of readers as a result of information contained in this text.

1 3 5 7 9 10 8 6 4 2

Contributors

Editorial Assistant

Dr Lucy Algeo

Expert Reviewers

Cardiology

Dr Kevin Fox

Consultant Cardiologist

Charing Cross Hospital

Dr James Ware

Clinical Fellow in Cardiology

Charing Cross Hospital

Respiratory

Dr Andrew Elkins

Consultant in Respiratory and Intensive Care Medicine

Princess Royal Hospital

Gastroenterology

Dr Nick Parnell

Consultant in Gastroenterology

Princess Royal Hospital

Artwork

Diagrams Pilar Zarate Ortiz, Cristina Koppel

THIS BOOK IS DEDICATED TO ALL THE EXCEPTIONAL TEACHERS I HAVE HAD THE PLEASURE TO ENCOUNTER, IN PARTICULAR:

Marlborough School, London

Miss Richards	Primary teacher

Hill House International School, London

Mr Brennan	English
Miss Capper	Science

Godolphin and Latymer School, London

Miss Babuta	English
Miss Hodgkins	Physical Education
Mrs Cockburn	History
Miss Clarke	French

Anglo-Colombian School, Bogota

Mr Peter Heffernan	Biology
Mr Keith Rigby	Chemistry

Imperial College School of Medicine, London

Dr Karim Meeran	Endocrinology
Prof. Diana Watt	Anatomy
Prof. Tony Firth	Anatomy
Dr Stan Head	Physiology
Prof. John Laycock	Endocrinology
Prof. Martyn Partridge	Respiratory Medicine
Dr Jacquie O'Keefe	Emergency Medicine
Dr Tim Orchard	Gastroenterology
Mr Barry Pareskeva	General Surgery
Mr Sanjay Purkayastha	General Surgery

New York University Medical Center

Dr Rodolfo Llinas	Neuroscience

My family, who are my life-long teachers

Claudia and Harry Koppel, Nelly Marquez and Alicia Duran de Koppel, whose autobiography "Somos Nuestra Memoria" inspired me to write this book.

CK

CONTENTS

List of Abbreviations

Ab	Antibody
ABC	Airway, breathing, circulation
ABG	Arterial blood gas
ABx	Antibiotic
ACE	Angiotensin converting enzyme
ACS	Acute coronary syndrome
ACTH	Adrenocorticotropic hormone
ADH	Anti-diuretic hormone
AF	Atrial fibrillation
ALP	Alkaline phosphatase
ALS	Advanced life support
ALT	Alanine transaminase
ANA	Anti-nuclear antibody
ANCA	Anti-neutrophil cytoplasmic antibody
ANP	Atrial natriuretic peptide
AP	Anterio-posterior
AR	Aortic regurgitation
ARDS	Acute respiratory distress syndrome
AS	Aortic stenosis
ASD	Atrial septal defect
ASO	Antistreptolysin O (Ab)
AST	Aspartate transaminase
ATII	Angiotensin II
AV	Atrioventricular/ arterio-venous
AVNRT	Atrio-ventricular nodal re-entrant tachycardia
AVRT	Atrio-ventricular re-entrant tachycardia
AXR	Abdominal X-ray
BBB	Bundle branch block
BCG	Bacille Calmette-Guérin
BHL	Bilateral hilar lymphadenopathy
BiPAP	Biphasic positive airway pressure

BM	Blood glucose (Boehringer-Mannheim made the machine)
BMI	Body mass index
BNP	Brain natriuretic peptide
BP	Blood pressure
Bx	Biopsy
Ca	Calcium
CA	Cancer
CABG	Coronary artery bypass graft
CBD	Common bile duct
CD	Crohn's disease
CF	Cystic fibrosis
CK	Creatine Kinase
CMV	Cytomegalovirus
CNS	Central nervous system
COC	Combined oral contraceptive
COPD	Chronic obstructive pulmonary disease
CP	Chest pain
CPAP	Continuous positive airway pressure
Creps	Crepitations
CRP	C reactive protein
CRT	Capillary refill time
CT	Computer tomography
CTPA	Computer tomography pulmonary angiogram
CVA	Cerebral vascular accident
CVP	Central venous pressure
CXR	Chest X-ray
D&V	Diarrhoea and vomiting
DIC	Disseminated intravascular coagulation
DVT	Deep vein thrombosis
ECG	Electrocardiogram
Echo	Echocardiogram
EDV	End diastolic volume

EMG	Electromyography
EMQ	Extended matching questions
ENT	Ear, nose and throat
ERCP	Endoscopic retrograde cholangiopancreatogram
esp.	Especially
ESR	Erythrocyte sedimentation rate
F	Female
FAP	Familial adenomatous polyposis coli
FBC	Full blood count
Fe	Iron
FEV_1	Forced expiratory volume at one second
FH	Family history
FNA	Fine needle aspirate
FOB	Faecal occult blood
FVC	Forced vital capacity
GFR	Glomerular filtration rate
GI	Gastrointestinal
GOJ	Gastro-oesophageal junction
GORD	Gastro-oesophageal reflux disease
GP	General practitioner
GTN	Glyceryl trinitrate
GU	Gastrourinary
Hb	Haemoglobin
HDL	High density lipoprotein
HIV	Human immunodeficiency virus
HNPCC	Hereditary non-polyposis colon cancer
HOCM	Hypertrophic obstructive cardiomyopathy
HPOA	Hypertrophic pulmonary osteoarthropathy
HRCT	High resolution computer tomography
HRT	Hormone replacement therapy
HS	Heart sound
HSV	Herpes simplex virus

HTN	Hypertension
IBD	Inflammatory bowel disease
IBS	Irritable bowel syndrome
ICP	Intracranial pressure
ICS	Inhaled corticosteroid
IHD	Ischaemic heart disease
im	Intramuscular
INR	International normalised ratio
IPPV	Intermittent positive pressure ventilation
iv	Intravenous
IVC	Inferior vena cava
IVDU	Intravenous drug user
Ix	Investigation
JVP	Jugular venous pressure
KUB	Kidneys, ureter, bladder (X-ray)
L	Left
LA	Left atrium
LABA	Long acting beta agonist
LAD	Left anterior descending (coronary artery)
LBBB	Left bundle branch block
LCA	Left coronary artery
LDH	Lactate dehydrogenase
LDL	Low density lipoprotein
LFTs	Liver function tests
LHF	Left heart failure
LIF	Left iliac fossa
LLQ	Left lower quadrant
LMW	Low molecular weight
LOS	Lower oesophageal sphincter
LRTI	Lower respiratory tract infection
LUQ	Left upper quadrant
LV	Left ventricle

LVF	Left ventricular failure
LVH	Left ventricular hypertrophy
M	Male
MAO	Monoamine oxidase
mc	Most common
MC&S	Microscopy, culture and sensitivity
mcc	Most common cause
MCQ	Multiple choice questions
MCV	Mean cell volume
MDI	Metered dose inhaler
MI	Myocardial infarction
MOD	Multi-organ dysfunction
MR	Mitral regurgitation
MRI	Magnetic resonance imaging
MS	Mitral stenosis
MSU	Midstream urine
N&V	Nausea and vomiting
NICE	National institute of clinical excellence
NSAID	Non-steroidal anti-inflammatory drug
NSTEMI	Non-ST-elevation myocardial infarction
OGD	Oesophagogastroduodenoscopy
OSCE	Objective structured clinical examination
PA	Posterio-anterior
PACES	Practical assessment of clinical examination
PCP	Pneumocystis carinii pneumonia
PCR	Polymerase chain reaction
PCV	Packed cell volume
PDA	Patent ductus arteriosus
PE	Pulmonary embolus
PEA	Pulseless electrical activity
PEEP	Positive end expiratory pressure
PEFR	Peak expiratory flow rate

PFO	Patent foramen ovale
Plt	Platelets
PMH	Past medical history
PND	Paroxysmal nocturnal dyspnoea
po	Orally
PPD	Purified protein derivative (tuberculin)
PPI	Proton pump inhibitor
PR	Per rectum
PRN	As required (pro re nata)
Pt	Patient
PTC	Percutaneous transhepatic cholangiogram
PTCS	Percutaneous transluminal coronary angioplasty
PTH	Parathyroid hormone
R	Right
r/o	Rule out
RA	Right atrium
RBBB	Right bundle branch block
RBC	Red blood cells
RCA	Right coronary artery
RDS	Respiratory distress syndrome
RF	Rheumatoid factor
RHF	Right heart failure
RIF	Right iliac fossa
RLQ	Right lower quadrant
RTA	Road traffic accident
RUQ	Right upper quadrant
RV	Right ventricle
Rx	Treatment
S1	First heart sound
S2	Second heart sound
S3	Third heart sound
S4	Fourth heart sound

SA	Sinoatrial
SALT	Speech and language therapy
Sats	Saturations
sc	Subcutaneous
SLE	Systemic lupus erythematous
SOB	Short of breath
SOBOE	Short of breath on exertion
STEMI	ST-elevation myocardial infarction
SVT	Supraventricular tachycardia
TB	Tuberculosis
TED	Thrombo-embolic deterrent (stockings)
TFTs	Thyroid function tests
TG	Triglyceride
TIA	Transient ischaemic attack
TNM	Tumour, nodes, metastases (for staging CA)
TOE	Transthoracic echocardiography
TPN	Total parenteral nutrition
U&E	Urea and electrolytes
UC	Ulcerative colitis
UFO	Unidentified flying object
URTI	Upper respiratory tract infection
US	Ultrasound
UTI	Urinary tract infection
V/Q	Ventilation/perfusion
VF	Ventricular fibrillation
VSD	Ventricular septal defect
VT	Ventricular tachycardia
VTE	Venous thrombotic event
w/	With
WBC	White blood cells
ZN	Ziehl-Neelsen (stain for TB)

INTRODUCTION

THE KEY TO PASSING FINALS

Most medical students will pass finals.

Medical finals are the last step in qualifying as a doctor. Their purpose is to confirm that students have reached a minimum standard before they're unleashed on the public. However, aiming for a higher standard makes you a better doctor and you will be more relaxed going into finals wondering if you'll get a distinction rather than worrying if you'll pass. As this is also the last formal exam before the MRCP, you will be judged on your result for the next couple of years. It is thus important in terms of future job prospects to do as well as possible to give yourself the option of choice in your future career.

Aim for honours.

CLASSIFY AND SIMPLIFY

You already know all you need to know to pass finals.

The amount of information you have to assimilate, understand and be able to regurgitate is formidable. To get the most out of your efforts, it pays to have a strategy. This book works in two ways. Firstly, it breaks down the work in a logical fashion. A large task becomes easier to undertake when it is broken down into manageable chunks of work. These have the benefit of giving you satisfaction and encouragement as they are completed. Secondly it simplifies that work by providing a structure that can be adapted to any medical problem. In the end, your workload is smaller and all you need are a few formulas to be able to answer almost any question that is thrown your way.

There is a limit to how much you are expected to know for finals. Understanding this is essential. It is far more valuable to have a solid understanding of the basics of all the subjects you may be tested on than to know a handful of subjects to specialists' depths. To really do well, you must do well across the board. Unfortunately, it is a known fact that students tend to shy away from the subjects they are weakest at and focus on their strengths. It might feel gratifying when revising, but unfortunately this will not help on the day when you are asked exactly what you avoided!

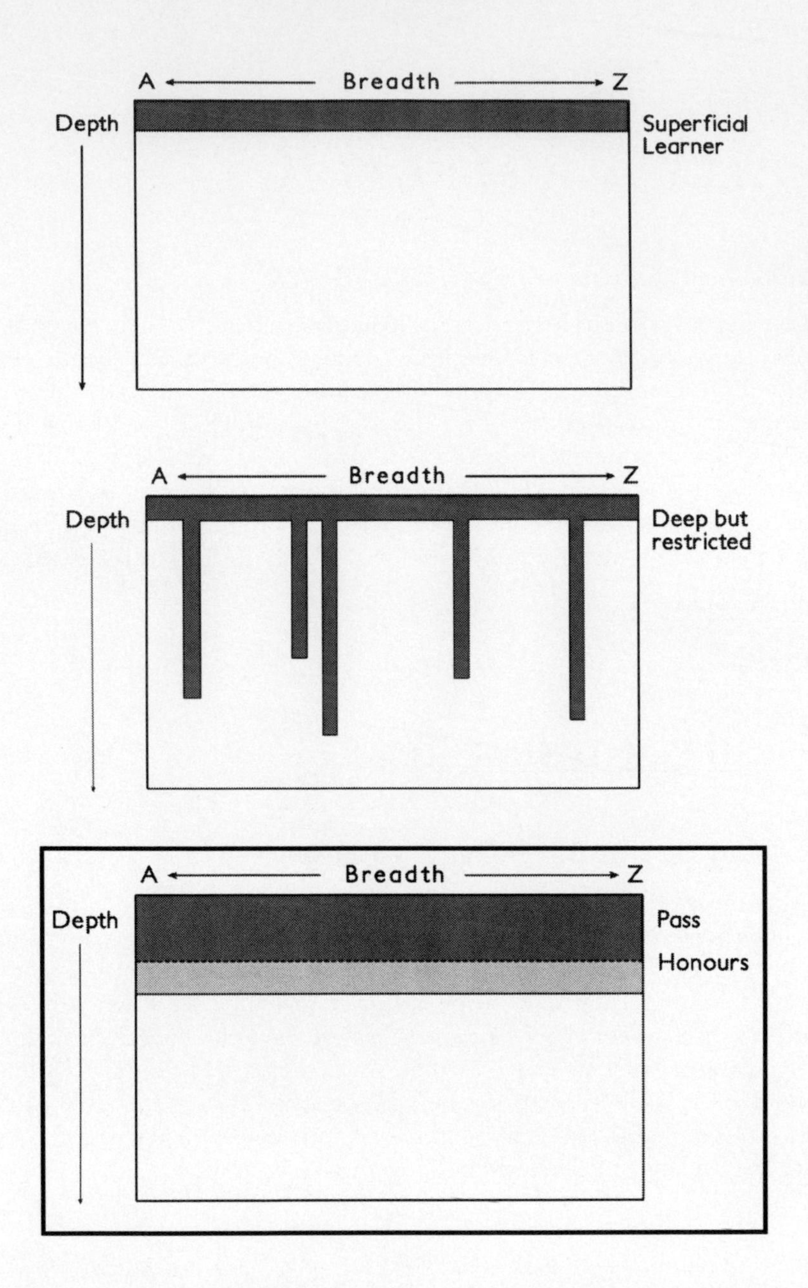

For each aspect of the exam it helps to have one, simple book to work through. This can act as a route map, helping you focus on what is important but also ensures you will not have missed anything major. You can always refer to bulkier texts to answer any remaining questions or to clarify certain points. One simple book plus a good MCQ/EMQ book along should be sufficient to form the foundations of your revision for each major subject (medicine, surgery, paediatrics, psychiatry, obstetrics and gynaecology).

Medicine simplified

The patient comes to you with symptoms of his pathology. The pathology itself is not usually physically visible, the notable exception being dermatology. Through your *history*, *examination* and *investigations* you diagnose the pathology by looking for evidence of

1. **The CAUSE of the pathology (aetiology)**
2. **The EFFECT of the pathology (symptoms and signs) including severity and complications**

Following your diagnosis, you *treat* the pathology, its causes and its effects.

The point of simplifying medicine like this is to give you a system to fall back on. Identifying the pathology allows you to move forward.

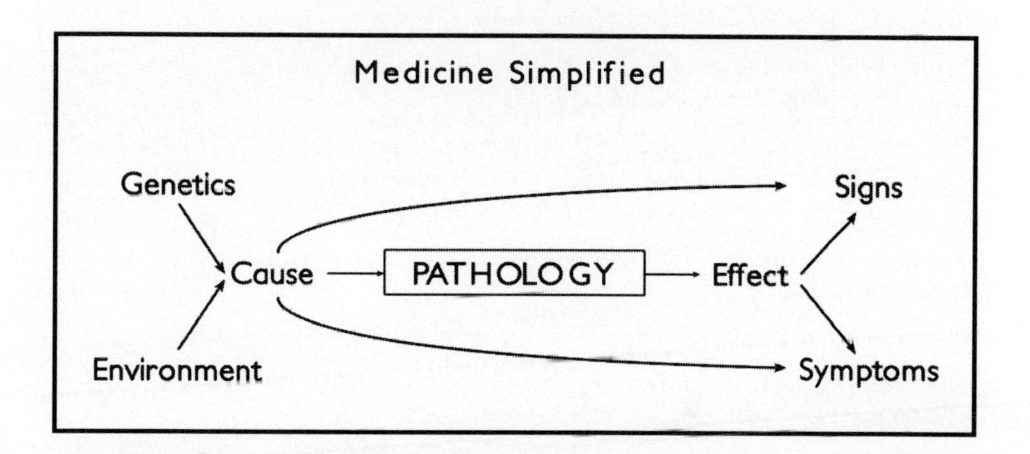

E.g. A patient presents to you with chest pain. The diagnosis of the pathology is reached by:

1. *History* of chest pain, nausea and sweating (effect); in a smoker with a strong family history of heart disease currently being treated for hypertension (cause).

2. *Examination* reveals he is anxious and clammy (effect); has raised BMI and xanthelasma (cause).

3. *Investigation* reveals ST elevation in the anterior leads on ECG, and a raised troponin (effect); normal fasting glucose but high lipids (cause).

The above evidence points to the pathology being myocardial infarction due to occlusion of his left coronary artery, most likely by thrombus. He is therefore treated with thrombolysis (treat pathology) oxygen and morphine (treat effect) and given lifestyle advice and started on a statin (treat cause).

Defining the problem

The patient gives you the history in layman's terms. Your job as a doctor is to listen carefully to the information you are given and ask the right questions to fill any gaps. You then "translate" the history into precise medical terms defining and simplifying the problem by condensing it to its essence. Creating this clear picture then allows you to communicate efficiently with other doctors. You can almost think of it as turning the features of this patient and his problems into an MCQ question.

E.g. 1.

You: Have you noticed any blood in the stool?

Patient: I don't really look at my stool, but come to think of it, yes, on occasions. It's fresh blood, sometimes on the paper.

You: How long has this been going on for and much would you say it is -a tinge, a teaspoon, a cup full, bucket loads?

Patient: Oh, no, just a tinge on the paper really, no more than a smidge on the stool. It doesn't really bother me -it's been going on for years. It gets better when I use a cream my GP gave me for my piles.

You: Any change in bowel habit, shortness of breath, fatigue, loss of appetite or weight loss?

Patient: No, I've been fine.

Now you can create the following picture:

A 52 year old man with known haemorrhoids, experiencing minimal fresh PR bleeding responsive to treatment with annusol. No constitutional symptoms or symptoms of anaemia. Impression: Haemorrhoids.

E.g. 2.

You: Have you noticed any blood in the stool?

Patient: I don't really look at my stool, but come to think of it, yes, on occasions. It's fresh blood, sometimes on the paper.

You: How long has this been going on for and how much would you say it is -a tinge, a teaspoon, a cup full, bucket loads?

Patient: Oh, it's a lot more than a tinge… it's hard to say, really, it's all mixed in and I don't pay it much attention, though it started about 2 months ago.

You: Any change in bowel habit, shortness of breath, fatigue, loss of appetite or weight loss?

Patient: Well, I've had this awful diarrhoea, though sometimes it swings the other way and I have to take laxatives the GP gave me. I've been off my food for the last couple of months which is why I suppose I've lost a stone or two, but that's good for me, isn't it? But I just feel exhausted all the time, lose my puff easily and my wife tells me I look pale.

Now you can create the following picture:

A 52 year old man with a two month history of PR bleeding, change in bowel habit and constitutional symptoms. He also has symptoms suggestive of anaemia. Impression: Bowel cancer.

Understanding is the key to remembering

As explained above, signs and symptoms have a *physical reason for being there* (even if we don't understand it!). By knowing the pathology and what is physically happening, you can explain its signs and symptoms. You could just memorise that aortic regurgitation gives you a large left ventricle, but if you *understand* that there is volume and pressure overload and its consequences, you will own the information and be able to work out all the other effects too. If you are ever stuck, go back to first principles and work it out.

EXAMS

1. Written papers

Obviously, a good breadth and depth of knowledge is essential. However, experience in answering MCQ/EMQs is invaluable. It is difficult to write these papers, therefore there are lots of very similar questions available in textbooks. They often anticipate hot topics and examiners' favourites and you will almost certainly find some of these questions come up in your own exams.

Work through questions under exam conditions if possible, especially if you are worried about time constraints. You will become familiar and more comfortable with the way questions are formulated. Don't look at the answer after every question; this decreases your learning and you might find that you are looking at the answers before even coming up with one yourself!

The key to answering MCQ/EMQs is to understand what the examiner is asking of you. A good start is to *read the question.* Silly as it sounds, it is not always done. Try to formulate an answer before looking at those provided. It is reassuring to find your answer amongst them. If it is not there, reconsider the question or look for the next best response. Answer every question as you go through, even if it is just a wild guess, but make a mark by those you are not sure of. Leave yourself some time to go back over these questions at the end – by that time you may have warmed up and the right answer will pop out at you.

That said, *do not change your answer* unless you can explain to yourself exactly why you are changing it. You are much more likely to change a correct answer you doubt to an incorrect one than the other way around (this has been statistically proven).

2. Practical assessment (OSCES/PACES)

There are three people in exam stations: *you, the patient and the examiner.* You have to understand each person's role.

You

The key to this part of the exam is to *play the game.* When you walk into the exam you are no longer a medical student. You are a doctor proving to your colleagues that you know how to deal with patients. Make the examiner feel that they would be comfortable working with you. You have to inspire trust by assuming a role of *confident humility*: confident that you can make a decent first assessment of a patient, but humble enough to know the limits of your own abilities and the value of experienced seniors' input.

This is your chance to shine. Think how you are going to show the examiners that you are able to talk to a patient, examine them thoroughly and come up with a sensible list of differentials, investigations and treatment. Finals are like taking your driving test. You know you can drive; now someone has to watch you do it. Every move must be witnessed to have been performed and should therefore be exaggerated.

CONFIDENT

Be confident (yes, even if you have to fake it!). This is as much a test of personality and stress management as it is of clinical skills. Use the minute before you enter each room to take a deep breath, relax and plan for the task ahead. Remind yourself the structure you worked on and the points you've tended to forget during practice.

Smile as you walk through the door.

1. Think like a doctor

Answer questions logically using a clear structure. If asked the causes of atrial fibrillation, just blurting out "pneumonia" doesn't show that you are thinking. Start with a general answer and move towards the specific -just *defining* atrial fibrillation could be a good start. There are tonnes of causes. Show the examiners you know this by breaking it down into cardiac, respiratory and metabolic causes, then name the most important ones in each; ischaemic heart disease, hypertension, pneumonia, sepsis, thyrotoxicosis and alcohol. End this short list with confidence. They can always ask you for more if they want it.

Remember that having a good structure is also your safety net: your mind may go blank when asked the causes of atrial fibrillation, but starting with the definition gives you time to think; saying you'd break it down into cardiac, respiratory etc. gives you yet more time. It is highly likely that during these seconds something will pop into your mind. And even if at the end of that you can't find the details, the examiner at least knows you have a sensible approach to medical problems. If you plan for every scenario, you should never look lost.

A smart candidate can lead the examiners questions in the direction of his choice by the way they answer the question. If you have particular knowledge that's relevant to the question, lead the examiner down that path and show him what you've got.

2. Talk like a doctor

You are talking to *colleagues*. You must learn to be eloquent (or once again, fake it for the duration of the exam!). Talk slow, think quick. Do not use abbreviations, especially if you don't know what they stand for as you'll probably be asked. Always use the appropriate medical terminology. This is easy to practice. Do not talk about "heart attacks, having a temperature and X-rays". The correct terms are myocardial infarction, pyrexia and radiographs. From now on, decide to actively think about how you are talking and make an effort to speak in medical terms (even with your friends) as you revise.

Make sure you can be heard by speaking with good volume. Be sure to finish your sentences with a clear, almost audible full stop. Insecurity is easily detected when answers trail off at the end or sound like a question.

Learning medicine is like mastering a language. Every word has its definition. Your job as a doctor is to translate what patients tell you into medical terms which allow you to define and thus simplify the problem then communicate it efficiently to others. Know your definitions and use them accurately.

3. Look like a doctor

Dress professionally in a clean suit. Remove visible piercings, hide those tattoos. If you want to be a doctor, you've got to look like one too. Your appearance (rightfully or wrongfully) provides a stream of clues to examiners as to who you are, and how you work. Once again, aim for confident humility. Leave the flashy shirt for another day; don't hide behind your hair.

HUMILITY

1. Admit your mistakes

If you say something wrong, *don't panic- you will only dig yourself deeper.* Recognising your mistake goes a long way to undoing the fact that you made it in the first place. Pause and take a step back. Inform the examiner and correct yourself, or if necessary ask if you can start again. (Remember it always helps to pause before answering any question as a preventative measure; avoid putting your foot in it at all costs). Also, each room is a separate exam. Do not go into the next scenario stressing about the last one; you have a new examiner and a clean white mark sheet. Focus on the task ahead and move on.

2. Recognise your limitations

Don't be afraid of reaching the end of your knowledge. If this is happening late in the exam and you are being asked difficult questions, it is a good sign. The examiners are pushing you to see how far you can go. If you really don't know the answer, say something *sensible* such as providing a source where you can find the answer, don't make things up. Useful phrases are "I would look it up in the British National Formulary" for drugs; "he was a turn of the century European physician" as a guess for dead and illustrious doctors/eponyms; "autosomal recessive" as a guess for most rare and terrible genetic diseases and you can always "ask for senior help". If all else fails, a simple "I don't know" allows the examiner to move on.

"I would ask for senior help" may well be the correct answer. Just as examiners would be reluctant to pass someone they felt knew too little or didn't have enough confidence, they are wary of the over-confident junior who thinks they are ready to put in a central line because they saw it once on ER. You are being examined on the knowledge and skills you need for Foundation Year 1, and knowing the appropriate time to call for help is an important a skill as any.

3. Don't fight the examiner

This is *not* the time for confrontation. Respect the examiners, they will be grading you.

By the time you get to the practical exam, you should have a solid foundation of medical knowledge and have a plan for every scenario that may come up. This book should provide you with the structure for most, but be prepared to formulate your own as necessary. Know your patients and know your examiners.

The patients

The patients that show up for finals share certain characteristics. They tend to be people who are reasonably well and have the time to attend. This limits the type of cases you get to mostly chronic, stable conditions with good signs in the elderly, such as aortic stenosis and COPD. You are extremely unlikely to get a young, newly diagnosed breast cancer patient -she will be too distraught with the diagnosis to be asked to come to an exam; likewise, a patient with acute heart failure or asthma will be too unwell. This results in a disproportionate emphasis on certain relatively rare conditions but also allows you to predict possible scenarios and focus on those you are more likely to encounter.

The other joy of having the patient there is that they will almost certainly be on your side, willing you on to answer the questions and spot the signs they are demonstrating. In that room, whatever the examiner is like, you definitely have one supporter.

The examiners

Broadly, there are two types of examiners:

1. The dove

Hopefully most of your examiners will fall into this category. They are rooting for you and are eager to see what you have to show them. They want you to pass. Don't disappoint them –shine! Listen to what they say and lookout for cues that they might be giving you. If you've forgotten to do something important, they might well nudge you in the right direction.

2. The hawk

These examiners are a little more difficult to deal with as they won't be looking to drag anything out of you; either you say it or you don't. Do not be intimidated by long silences. They do not mean you are wrong. Do not change your answer in an attempt to cover up the silence. Let the examiner be the one to break it unless you know exactly what you are talking about. This is just his way of discerning your degree of confidence in an answer. The way you talk to these examiners is key. It is even more important to complete your sentences and end clearly; otherwise all they will do is stare at you indefinitely and watch you squirm as you dig yourself deeper.

PRACTICE

Practice makes perfect. It really does. Divide your practice into three types:

1. On friends and family

This type of practice should be done early on in revision. Go over *each and every* examination until you can do it in your sleep. This means you can actually look for the sign instead of thinking what comes next. By continuously examining normal people in a friendly, relaxed environment, you will be able to build a routine you feel confident with, is comfortable for the patient and looks *slick* to the examiners. Work on creating a tempo and being thorough. No movement should be wasted, you should also be able to explain and justify your methods. Make sure to use an adaptable structure, e.g. only move on to shifting dullness if you percuss the abdomen and suspect ascites, there is no point doing it otherwise.

Having this routine under your belt not only gives you the confidence that you'll be able to perform under the extra pressure of exam conditions; it frees your mind up to analyse what you're seeing and hearing. It means you won't miss anything out, and convinces the examiner that you've done this before. If you examine quickly you also leave more time for questions and extra bonus points.

2. On patients

The aim of examination is to pick up signs. It is paramount to attempt to have seen all the signs you might be expected to spot in the examination. You are more likely to be successful if you know what you're looking for and have acquired a technique that will enable you to find it. Find patients on the wards to examine. Keep in mind that as the exam approaches, the patients will have been seen by countless students and might not consent to examination. Once again, start early. You might be surprised to get a patient you already know in your exam (don't tell the examiners but don't deny it if asked!)

3. With your tutor

Sort out tutor groups early. It's great if you can find someone reliable who has a plan to go over the most important topics with you. Ideally, you should arrange several types of teaching with different individuals. Hospitals are full of doctors – if you don't get on with your own tutor for whatever reason, then go in search of your own teaching. There are a surprisingly large number of people out there willing to help. If you don't ask, you don't get.

General grilling is very useful in getting students to lose their fear of the examiner, use medical terminology and express themselves logically under pressure. Despite this being a very effective form of learning, students dislike it as it puts them on the spot. Learn to deal with it; it'll be much easier on the day if you've been through worse before. You should also see patients with your tutor as they will fine-tune your examination technique and provide nuggets of advice you often don't find in textbooks. Take every opportunity to present the patient to them. Once again, this puts you on the spot (good exam experience) and helps you learn to create logical, concise summaries. This is often a weak skill for most candidates so an eloquent summary should make you stand out.

HOW TO USE THIS BOOK

Although this book is aimed at final year medical students, it may prove useful to students in earlier years. However, it does assume a certain level of knowledge. The aim is to structure the knowledge you already have, integrating clinical experience with the theory studied in pre-clinical years.

Communication skills must not be neglected as it is an ingredient of the holistic, patient-friendly approach which forms part of the treatment of any disease. Also, in exams there are usually specific marks for this. As modern medical schools place a large emphasis on communication skills and this subject is particularly well taught, this book centers on structure and demonstration of medical knowledge and understanding.

You should aim for honours, therefore the level of this book is that required to accomplish this.

Each chapter is devoted to a system and is divided into 5 sections:

History

Examination

Presentation

Investigations

Pathology

The key concepts behind each of these sections are explained below. The cause/effect simplification for collecting evidence of the pathology has been followed throughout and is worth repeating:

The patient comes to you with symptoms of his pathology. Through your *history*, *examination* and *investigations* you diagnose the pathology by looking for evidence of

1. **The CAUSE of the pathology (aetiology)**
2. **The EFFECT of the pathology (symptoms and signs) including severity and complications**

Following your diagnosis, you *treat* the pathology, its causes and its effects.

HISTORY *Key questions and symptoms*

Medical students are usually pretty good at following the history template:

Presenting complaint
History of presenting complaint
Systems review
Past medical history
Drug history
Family history
Social history, including "Ideas, Concerns and Expectations"

This section provides a template of key questions addressing each **cause** (risk factors) and **effect** (symptoms) of pathologies for a particular system. You should still follow the traditional history template, but have these in the back of your mind. These questions can also be used for systems review and if answered in the positive, you can ask further questions to flesh out the details

Tips for history taking

1. Know what is important and bring it to the front

Unfortunately, key questions are often missed out of the HPC. Important facts that fall under other categories are also frequently forgotten (e.g. cardiac risk factors are left scattered and hidden in the depths of family, past medical and social history).

2. Be flexible

Taking the history itself is **not** a linear progression through a template. Rather listen out for clues patients give, follow them up and then return to the initial thread. On the following page, an example for a history of chest pain is given.

3. Construct a timeline

It is useful to be able to draw a **timeline** of symptoms.
E.g. A patient presenting with SOB

SOB started six days ago, followed by a productive cough. Over the last four days SOB has been increasing and cough has been productive of initially yellow, now green sputum (no blood). Two days ago the GP started amoxicillin, with no response.

4. You might not be the first doctor they've consulted

Try to find out exactly what **investigations and treatments** the patient has already had.

5. Quantify as much as possible

Objective measuring allows you to track the patient's progress. Severity scales are provided in this book where practical and validated. E.g. New York Heart Association Class for heart failure.

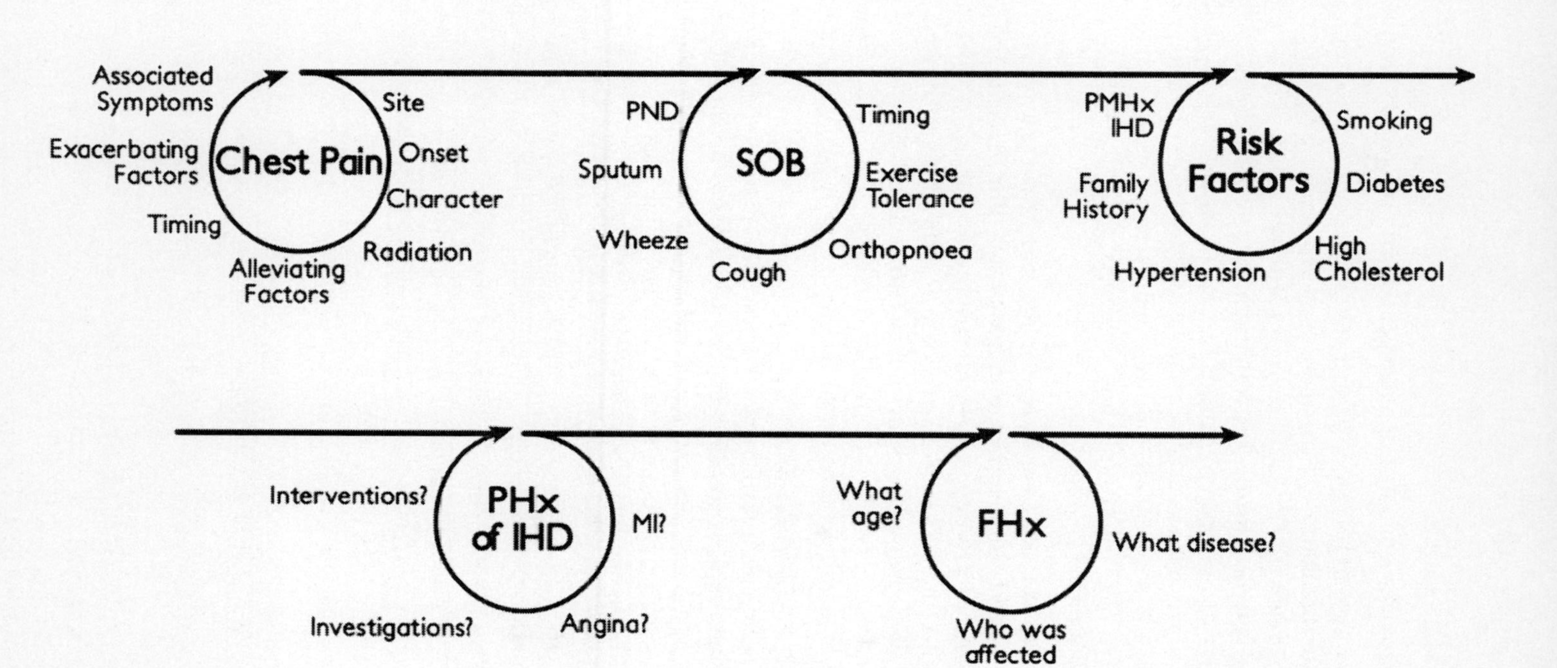
Chest Pain
Associated Symptoms
Exacerbating Factors
Timing
Alleviating Factors
Radiation
Character
Onset
Site
SOB
PND
Sputum
Wheeze
Cough
Orthopnoea
Exercise Tolerance
Timing
Risk Factors
PMHx IHD
Family History
Hypertension
High Cholesterol
Diabetes
Smoking
PHx of IHD
Interventions?
Investigations?
Angina?
MI?
FHx
What age?
Who was affected
What disease?

EXAMINATION

This section provides a template for examination of each particular system.

Have your own system based on the accepted sequence of inspection, palpation, percussion and auscultation. Examiners expect a certain progression although they accept that each person has their own particular adaptations to the general formula. Find out what works for you and have a reliable method for finding each sign. You can use any method you like as long as you can justify it, for example, you may want to auscultate the front of the chest before going on to palpate, percuss and auscultate the back, simply to prevent an elderly patient from having to move forwards and backwards all the time. Always be considerate, *do not cause the patient pain.*

Remember there is a concrete cause for the signs that you pick up. Try to imagine what's physically happening in the body. For example, colon cancer and deep palpation of the abdomen: picture the firm mass of the tumour and feel for it, under the skin amongst the soft intestines. You can practice this hiding a ping pong ball under a soft pillow and feeling for it. If you prod the pillow here and there, you'll never find it. If you feel the pillow systematically with your whole hand, pressing down far enough with a rolling motion, you're much more likely to find it.

Examination combines the ability to take in what is jumping out at you with the skill of searching for what you suspect. So, inspection is not just about *seeing* the obvious signs in a patient, is also about *looking* for more subtle clues that your medical knowledge tells you might be there. The same idea holds for palpation, percussion and auscultation. Examination is a thinking process and not about mindlessly carrying out a set of independent movements. Form an idea of what is happening as you go along and spot the connection between signs as this will point to the underlying pathology. Likewise, be surprised and reevaluate your diagnosis if things aren't adding up.

Practice makes perfect. Examination should be slick and second nature so that you can concentrate on what you are finding rather than worrying about which part of the examination comes next.

Examination

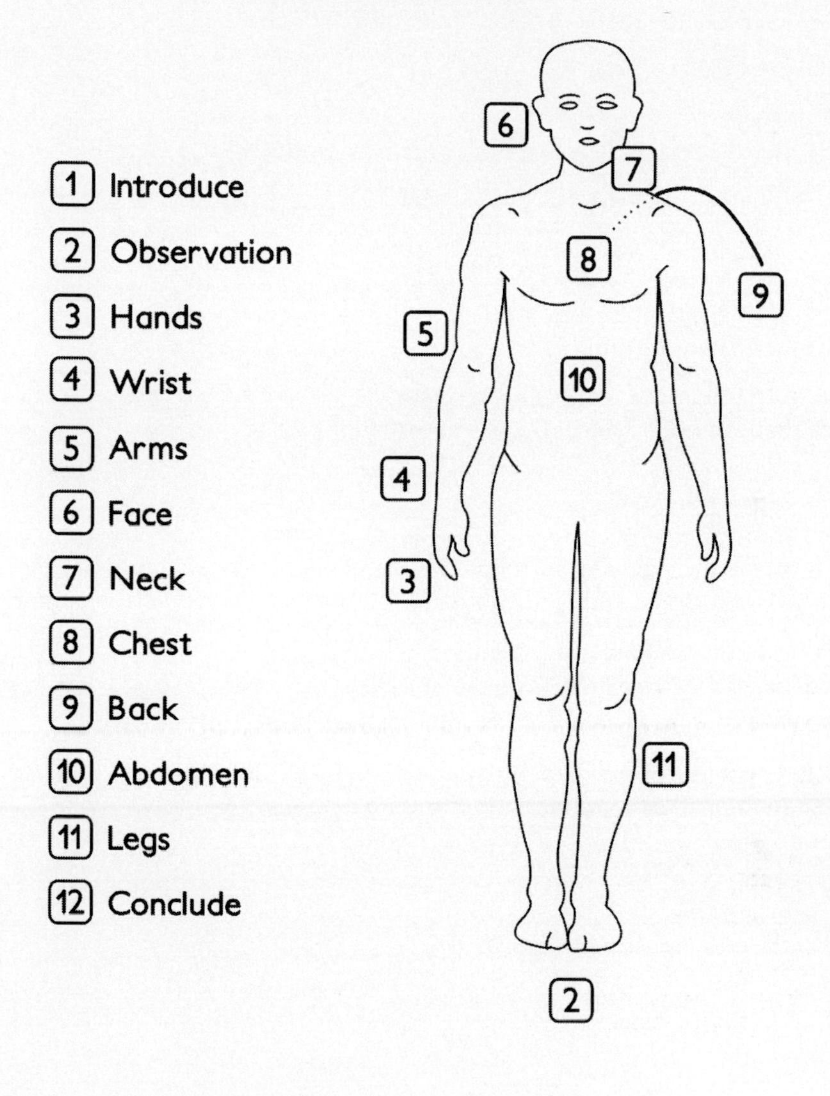

Introduce

Always introduce yourself, state your name and role. Ask for consent before starting your examination.

Think "HELP"

Hello

Expose

Lighting

Position

Name/age/occupation

Don't forget this one or you'll be stuck when you start to present them and realise that you don't have this basic information.

Observe

First observe the patient's surroundings, then the patient themselves. Do this from the foot of the bed to demonstrate to the examiner this is what you are doing.

Hands/ Wrist/ Arms/ Face / Neck/ Chest/ Back/ Abdomen

This template is followed for each system, which allows you to integrate the examination of all systems into one if necessary.

Legs

Start with the hands *and end with the legs!*

Legs are often forgotten and are important, if only to look for oedema which is relevant to every system.

""To complete my examination, I would like…"

This includes simple bedside tests, rather than anything more complex. For example, dipsticking urine, but not sending urine for an MSU.

Turn and talk to the examiner. Following this they will ask you to present your findings.

Presentation

This section is the least practiced and the one students struggle most with. Get this right and you're sure to impress. Thank the patient and turn to face your examiner. Take your stethoscope from around your neck, fold it up and stand with your hands behind your back. This gives you an extra five seconds to collect your thoughts. Speak with confidence and state what you have found in a logical order using the appropriate terminology. Do not berate yourself or make excuses. Look straight at the examiner. There should be no need to look back at the patient; you have finished your examination. Shorten long lists of negatives e.g. "there are no stigmata of chronic disease" or "I found no positive signs". Give your impression and a concrete diagnosis with differentials if you can. If you are feeling particularly brave you can follow this with the investigations you would like to carry out on this patient.

Investigation

This section goes over simple investigations you could be asked to interpret for each system. Once again, be systematic. Begin by stating the patient's identity (e.g. Mr George Smith), the investigation you are looking at (e.g. a PA chest radiograph) and adequacy of that investigation (e.g. a rotated film). This gives you time to look for abnormalities. Start with the template but don't be surprised if the examiner wants you to cut to the chase due to time constraints. If there is an obvious abnormality, say so after identifying the investigation. Remember that you need not be at radiologist/cardiologist level when interpreting radiographs/ECGs. If you can spot the abnormality, you can then show it to your senior or a specialist who can worry about the details.

It is helpful to understand that there is a physical, concrete reason behind what you see on a radiograph/ECG. Imagining what you're actually looking at in the human body, the pathology and *why* it looks like that allows you to understand and therefore interpret the results.

There are certain ways you expect pathology to show up, e.g. tracheas *shift* (either pushed or pulled by a lesion), costophrenic angles become *blunt*, QRS complexes are either *wide* or *narrow*, ST segments become *depressed* or *elevated*, while T waves *invert*. Focus on each element of the investigation in turn, see how they are behaving and look out specifically for those changes that commonly occur.

PATHOLOGY

This section goes through the different diseases in each system. They have been simplified and structured to provide the essence of each disease. Medicine involves a lot of lists which unfortunately have to be learned. Structuring each list allows more to be remembered with less effort; if you understand the disease and can remember the headings of your structure, you can work out the entire list.

Definition

This is ever so important. Know what you are talking about. If asked about any disease, defining it is a good place to start. It also buys you time. Remember, you can be quite general and simplistic yet still sound clever, e.g. COPD = "A chronic, progressive respiratory disorder most commonly seen in smokers characterised by airway obstruction with little reversibility".

Description

It is useful to know a brief summary of the epidemiology of a particular disease. Don't bother remembering the exact incidence of a disease; it's much more useful to know whether it is relatively common or incredibly rare. Most diseases can happen in any age group, sex or ethnicity. These are just guides to provide a picture of the typical patient.

Aetiology

This is the cause (or causes) of the pathology, and the risk factors should be screened for in your history if you're suspecting it.

Pathophysiology

These are the changes in mechanical, physical, and biochemical functions associated with or resulting from disease. This is important as it allows you to understand the concrete changes that are occurring in the body. The causes and the effects of these changes are what you are looking out for in the history, examination, then investigating and subsequently treating. If you understand how the disease behaves, you should be able to work everything else out.

History and Examination

This section is divided into causes and effects. Effects are generally more obvious as they are what the patient presents with. In most cases causes and risk factors have been discussed under "aetiology". You should remember to screen for them too. Unlike many medical texts, the segregation into signs and symptoms is not always made because this structure is not as helpful for remembering them, neither is it always clear what category they fall into, for example jaundice –who noticed it first, the doctor or the patient? What is important is that they are manifestations of the same pathology.

Investigations

Once again, these are looking for causes and effects of the pathology. Imaging allows the pathology itself to be visualised. The aim is to diagnose the disorder, exclude differentials, assess severity and look for complications. Essential investigations for each disease are highlighted in bold in the text (for example ECG and troponin in suspected MI).

Simple Bedside tests such as PEFR, ECG
Blood Haematology/ Biochemistry/ Microbiology
Imaging X-ray/ US/ CT/ MRI
Invasive e.g. endoscopy, angiogram
Special e.g. lung function tests

E.g. For atrial fibrillation:

Simple **ECG**
Effect -irregularly irregular rhythm, no P waves

Bloods **FBC, cardiac enzymes, TFTs**
Cause -infection, MI, hyperthyroidism

Imaging **Echocardiogram**
Cause and *effect* -heart failure

Treatment

You should consider the treatment of the pathology including its causes and effects, as well as complications. You can't go wrong using the buzzword "multidisciplinary team" especially for diseases like cancer.

Conservative e.g. physiotherapy, dietary changes

Medical i.e. drugs

Interventional "Interventional" is used rather than "surgical" due to the increased use of interventional radiology, such as angioplasty

Prognosis

Have some idea of the natural history of the disease, response to treatment and complications.

In every case, structure is more important than having a list.

If nothing else, it makes lists easier to remember.

More importantly, it demonstrates logical thinking.

CARDIOLOGY

CARDIOLOGY

HISTORY

CAUSE

Cardiovascular risk factors

1. Hypertension
2. Smoking
3. Cholesterol
4. Diabetes
5. Family history
6. Known ischaemic heart disease

EFFECT

Symptoms and severity

1. Chest pain

"SOCRATES"	*Textbook angina*
Site	Central
Onset	Sudden
Character	Crushing
Radiation	Neck/jaw
Associated factors	N&V, sweating
Timing	On exertion
Exacerbating/alleviating factors	Exercise/ GTN
Severity (from 1 to 10)	10 -worst pain ever

How severe is their angina?

Unstable angina

Symptoms at rest

Symptoms worsening over < 1 month (crescendo)

Canadian Cardiovascular Society Grade	
I	Symptoms on strenuous activity
II	Symptoms on moderate activity (stairs)
III	Symptoms on mild activity (flat)
IV	Symptoms at rest (unstable angina)

2. Evidence of heart failure

Shortness of breath: Determine "Exercise tolerance"

SOB at rest?

Walking distance before stopping for breath?

Flights of stairs/number of steps before stopping for breath?

New York Heart Association Class	
I	Asymptomatic during ordinary activities
II	Symptoms during ordinary activities
III	Symptoms during less than ordinary activities
IV	Symptoms at rest

Orthopnoea

Can they lie flat?

How many pillows do they sleep with at night?

Paroxysmal nocturnal dyspnoea

Do they wake up breathless at night?

Palpitations

Can they feel their heart beating in their chest?

Is it fast/slow, regular/irregular

Asking patients to tap out the rhythm can be useful

Associated symptoms

Swelling of ankles

EXAMINATION

Introduce

Name/age/occupation

Observe

1. **Surroundings** (Oxygen cylinder, walking stick)
2. **BMI** (do not call your patient "fat")
3. **Shortness of breath**

Hands Clubbing, splinter haemorrhages, peripheral cyanosis,
Capillary refill time

Wrist **Pulse**: Regular/irregular

Arms
1. **Collapsing pulse**:
2. **Blood pressure**:
   "I would like to take the blood pressure at this point"
   Hyper/hypotension
   Wide pulse pressure -Aortic regurgitation
   Narrow pulse pressure -Aortic stenosis

Face
1. **Eyes**: Anaemia (you only need to look in <u>one</u> eye)
   Xanthelasma
   Lipid arcus
2. **Mouth**: Central cyanosis

Neck
1. **Carotid pulse**
2. **JVP**
3. **Corrigan's sign**

Chest
1. **Observe** carefully for scars

2. **Palpate**

1. The apex beat

Displaced	Dilated left ventricle or mediastinal shift
Heaving	LVH
Tapping	Palpable 1st HS of MS

2. Heave (right sternal edge) Pulmonary hypertension
3. Thrills (over valves) Palpable murmur

3. **Auscultate**

1. All four valve areas for heart sounds ±added sounds
2. Apex on expiration and left side

   Radiation to axilla?
3. Right sternal edge on expiration and leaning forward.

   Radiation to neck?
4. Carotids for bruit

If you hear a murmur, ***LISTEN TO IT CAREFULLY***

1. **Timing**	Systolic/diastolic
2. **Location**	Aortic area/ apex
3. **Accentuation**	Inspiration/ expiration
4. **Position**	Leaning forward/ left lateral
5. **Radiation**	Axilla/ neck

6. **Grade**

I Very soft (experts only!)

II Soft

III Clearly audible, no thrill

IV Clearly audible + thrill

V Loud + thrill

VI Heard without stethoscope + thrill

In finals, murmurs heard will tend to be III - IV, so the important part is deciding whether a thrill is present or not.

Back Auscultate lung bases for bibasal crepitations (LHF)

Look for sacral oedema (RHF)

Abdo Hepatomegaly (RHF)

Legs Oedema, vein harvesting scars for CABG

Xanthoma of Achilles tendon

Peripheral (foot) pulses/ sensation as a quick screen for diabetes

"To complete my examination, I would like…"

Temperature/ BM/ urine dipstick/ 12 lead ECG

PRESENTATION

1. Name/ age/ occupation

2. Appearance Surroundings/ BMI/ SOB

e.g. "The patient had a high body mass index and was short of breath at rest and sitting upright. He uses a walking stick."

3. Peripheral signs: Hands, wrist, arms, face and neck

Positives & relevant negatives:

e.g. "There was no clubbing, his pulse was of good volume and irregularly irregular. His blood pressure showed a wide pulse pressure. He had no signs of anaemia or cyanosis. Xanthelasma was present around his eyes. His jugular venous pressure was raised 6cm above the sternal notch. Corrigan's sign was also present."

4. Chest: Observation, palpation, auscultation

e.g. "On examination of his chest, there were no scars, the apex beat was displaced to the mid axillary line, 6th intercostal space. I felt no thrills or parasternal heave. On auscultation, I heard a grade a grade III diastolic murmur, best heard at the left sternal edge on expiration with the patient leaning forward. This radiated to the axilla."

5. Back and legs

Right heart failure: leg oedema. Left heart failure: bibasal crepitations

e.g. "This patient had pitting oedema to his knees and bibasal crepitations on auscultation of his lung bases."

6. Summary

Put everything together with positives and relevant negatives. Don't stop at the diagnosis. Try to comment on cause and effect (severity/ presence of heart failure) while classifying and quantifying what you can.

e.g. "In conclusion, Mr Smith is a 79 year old, gentleman with cardiovascular risk factors who has evidence of atrial fibrillation and severe aortic regurgitation which has resulted in New York Heart Association Class IV congestive cardiac failure. My next step would be to investigate with echocardiography"

Explanation: The aortic regurgitation is severe because of the peripheral signs (Corrigan's and a wide pulse pressure) and heart failure, NOT because of the grade of the murmur. It is congestive cardiac failure because there is evidence of both left (SOB + bibasal creps) and right heart failure (leg oedema and ↑ JVP). He is in NYHA Class IV heart failure as he is short of breath at rest.

Note: Be wary of signs that don't add up, e.g. wide pulse pressure and Corrigan's sign with the murmur of aortic stenosis points to mixed valve disease. Aortic regurgitation can be hard to hear when the harsh murmur of aortic stenosis is present. The dominant lesion can be determined by the pulse pressure.

Investigation

ECG

ECG interpretation can be as easy or as difficult as you want it to be. A basic level is expected of you but cardiologists can read futures in these moving lines. The more time you spend interpreting and understanding ECGs the more you will get out of them.

See http://www.thinkingmedicine.com/elearning/electro/ for an interactive tutorial.

Describe	Identify investigation and patient. "This is an electrocardiogram of (whoever) taken (whenever)"	
Adequacy	Gain, all leads present	
Rate	Tachycardic/ bradycardic	
Rhythm	Regular + P wave	=Sinus rhythm
	Irregularly irregular	=Atrial fibrillation (mcc)
	Regularly irregular	=Heart block
	Look out for ectopics (atrial -narrow/ ventricular -wide)	
Axis	Left/ right axis deviation	
P waves	Present	=Sinus rhythm
	Not present	=Atrial fibrillation (mcc)
PR interval	Normal	3-5 small squares from start of P to start of R
	Long	=Heart block
	Short	=Wolff-Parkinson-White
Q	>2mm	=Previous full-thickness MI

R	Progression	Poor progression V1-V6 = previous ischaemia
QRS	Wide	= Ventricular arrhythmia/ Bundle branch block (>0.12ms/ 3 small squares)
	Narrow	=Supraventricular rhythm
ST	Elevation	=Acute coronary syndrome (STEMI)
	Depression	=Acute coronary syndrome (possible NSTEMI)
T	Inversion	=Acute coronary syndrome (possible NSTEMI)
	Saddle	=Pericarditis
	Peaked	=Hyperkalaemia

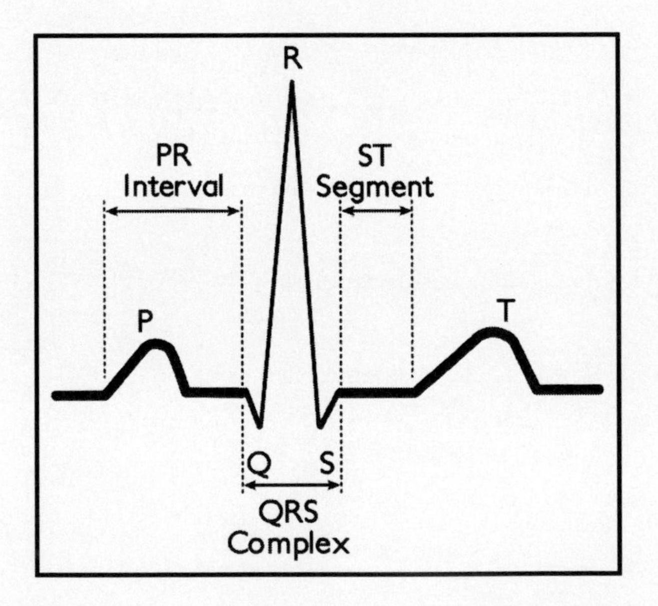

For the Thinking Medicine online ECG tutorial go to:

http://www.thinkingmedicine.com/elearning/electro/

PATHOLOGY

Arrhythmias

Do not let yourself be scared by arrhythmias. They are often poorly understood as they are poorly taught. Their confusing nomenclature compounds the frustration. This book aims to give a general overview of the key concepts that will allow you to have a general understanding of arrhythmias. You are unlikely to encounter anything other than atrial fibrillation in the practical exam, (unless they show you an ECG) and written exams are unlikely to demand the finer details. *Atrial fibrillation is the one rhythm you should know well.*

Definition: Abnormality of cardiac rhythm.

Arrhythmias can be categorised by their location and their rhythm.

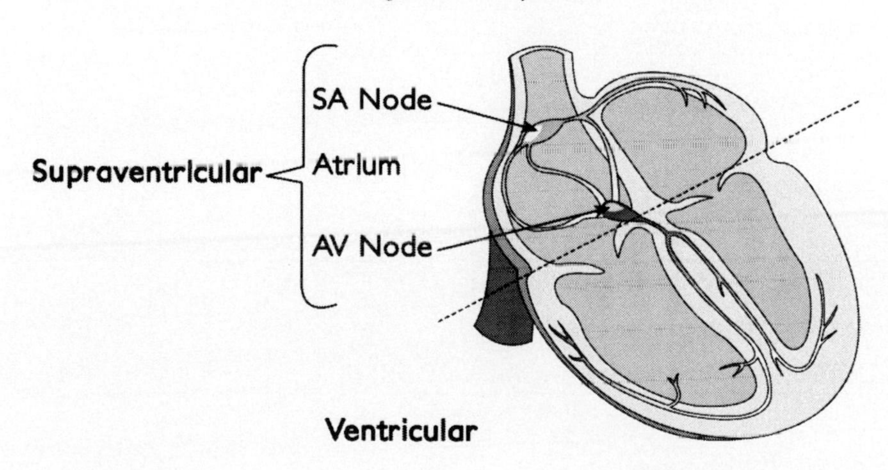

1. Location

Supraventricular

These originate from anywhere above the His-Purkinje system, resulting in **narrow** QRS complexes; (unless there is co-existing bundle branch block which will widen all the complexes but the origin is still atrial).

Sinus	Arising from the SA node e.g. Sinus rhythm
Atrial	Arising from anywhere else in the atria e.g. Atrial fibrillation/flutter
Junctional/nodal	Arising from the AV node e.g. AVNRT/AVRT - sometimes called "SVTs"

Understand that the term SVT can refer to two different things:

1. To mean "*any* rhythm originating above the AV node" (i.e. sinus, atrial or junctional) or

2. Misused to mean a *specific* rhythm originating close to/at the AV node (i.e. AVNRT/AVRTs=junctional=nodal)

Ventricular

These originate from anywhere in the ventricles, resulting in **broad** QRS complexes.

Ventricular tachycardia
Ventricular fibrillation

Correlating the ECG to these arrhythmias is helpful

P wave	**= SA node functioning**
Narrow QRS	**= Supraventricular rhythm**
Broad QRS	**= Ventricular rhythm (or BBB)**

2. <u>Rhythm</u>

Though it is useful to think of arrhythmias in terms of location as above since this correlates closely with the ECG, they can also be classified in terms of rhythm as follows.

Bradyarrhythmias <60bpm

Supraventricular	Sinus bradycardia Atrioventricular block (heart block)
Ventricular	Bundle branch block *(not technically bradycardia)*

Tachyarrhythmias >100bpm

Atrial	Sinus tachycardia
	Atrial flutter
	Atrial fibrillation*
Junctional	AVNRT/AVRT
Ventricular	Ventricular tachycardia
	Ventricular fibrillation

BRADYARRHYTHMIAS

Sinus bradycardia

= slow rate <60bpm, but normal conduction from SA node

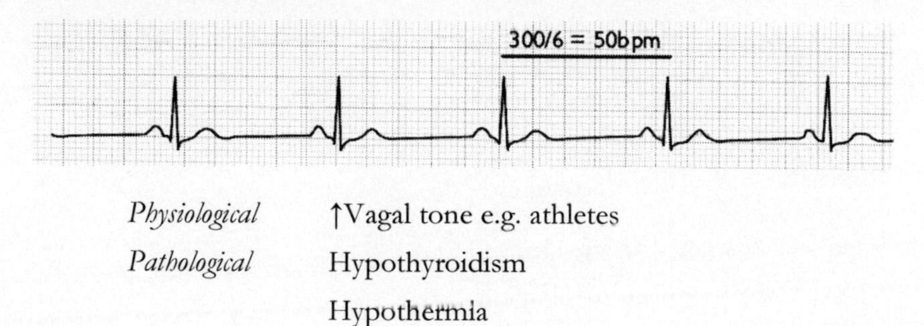

Physiological	↑Vagal tone e.g. athletes
Pathological	Hypothyroidism
	Hypothermia
	↑ Intracranial pressure
	Drugs e.g. β-blockers, digoxin

Bradycardia may lead to hypotension and require treatment with atropine or pacing

Atrioventricular block (Heart block) = delay in transmission of impulses from atria to ventricles. The QRS complex is normal as it is supraventricular in origin.

1st degree Prolonged PR interval (>0.2 ms / 5 small squares)

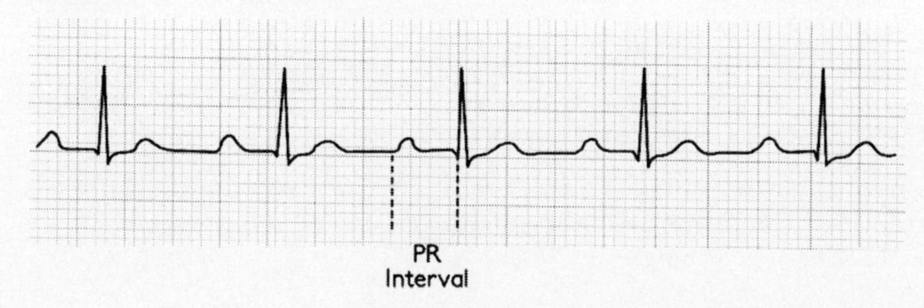

2nd degree **Mobitz I /Wenckebach**

=Cycles of successive prolongation of PR interval then dropped QRS:

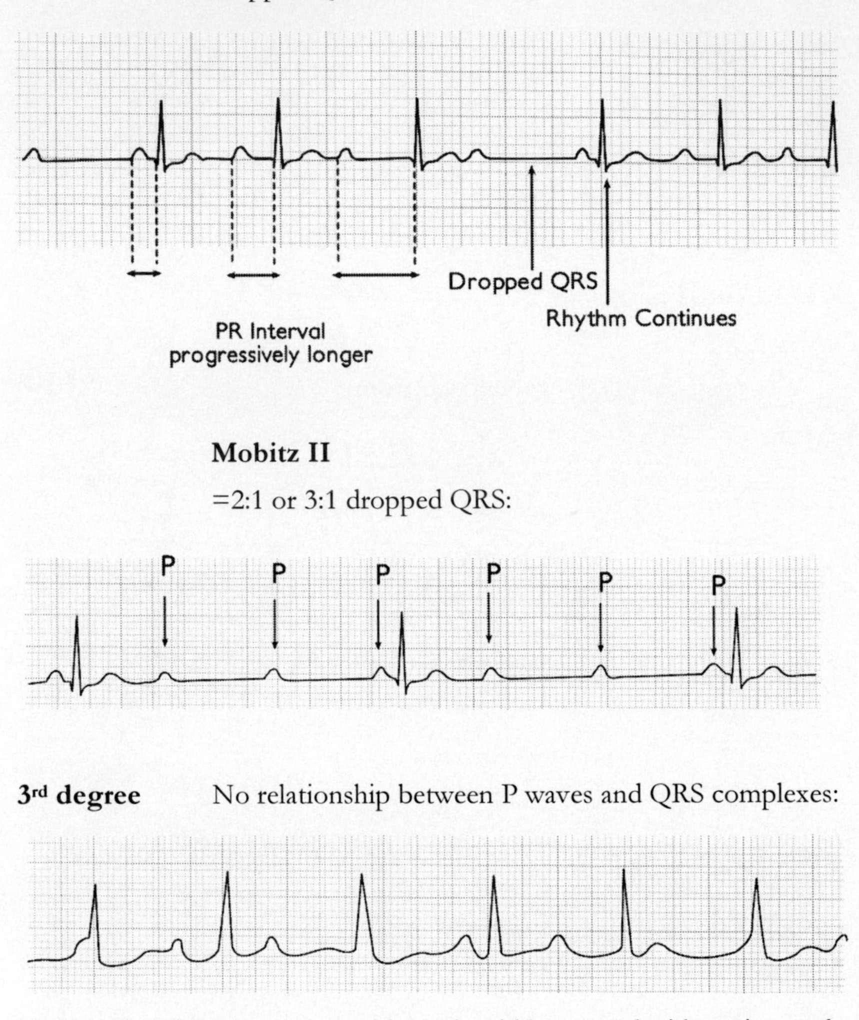

Mobitz II

=2:1 or 3:1 dropped QRS:

3rd degree No relationship between P waves and QRS complexes:

Mobitz II and 3rd degree heart block should be treated with pacing as there is an increased risk of asystole.

Bundle branch block (BBB) = delayed ventricular conduction results in broad QRS complexes -note that the *origin* of the complex is still atrial.

BBB is **not** a bradycardia, but it doesn't really fit in anywhere else. If present, the rest of the complex cannot be interpreted, so ST changes etc. in the presence of BBB are meaningless.

BBB on top of a supraventricular tachycardia may resemble ventricular tachycardia as the broad QRSs (of *atrial* origin though!) are paired with a fast rate.

The easiest way to determine BBB is to remember the mnemonic:

MaRRoW = *Right BBB*

A broad M-shaped QRS in V1 progressing to a W-shaped QRS in V6

WiLLiaM =*Left BBB*

A broad W-shaped QRS in V1 progressing to an M-shaped QRS in V6

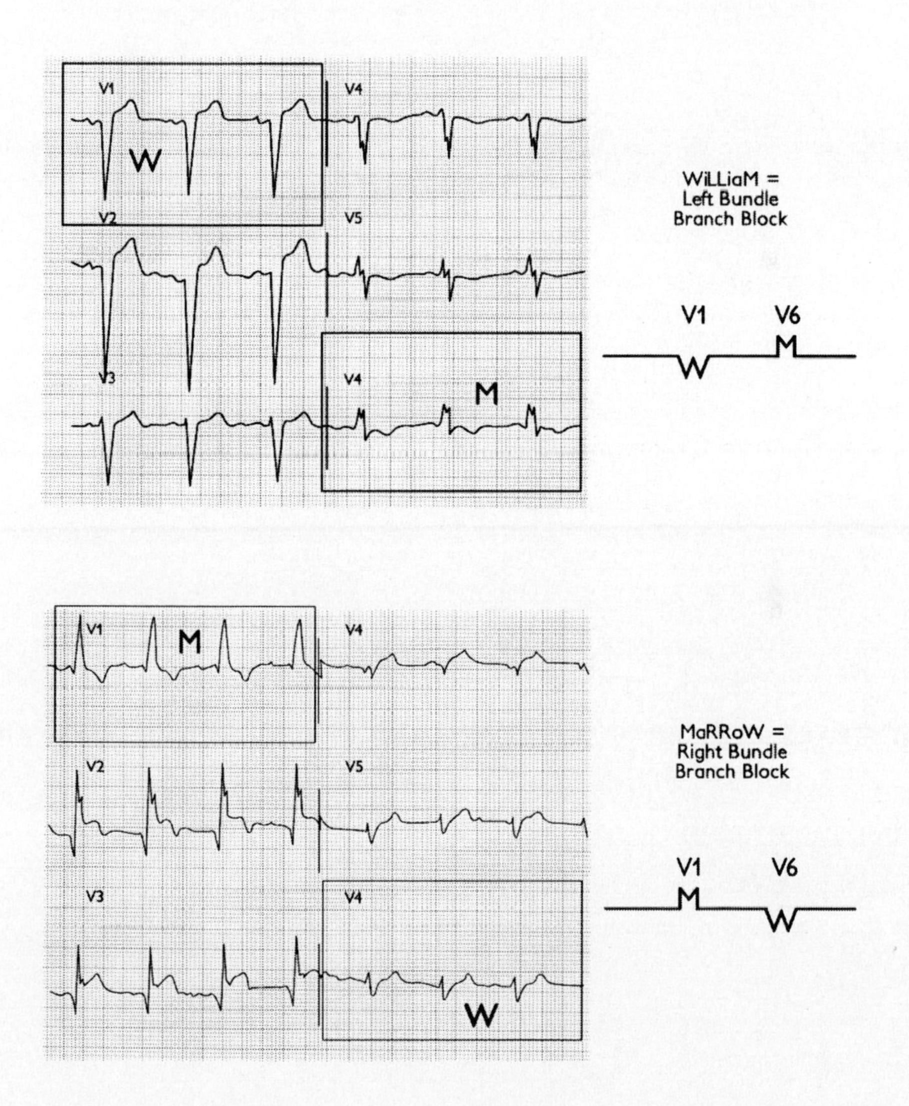

In addition:

Escape rhythms may be seen in bradycardias. This is a protective mechanism where the next most active intrinsic cardiac pacemaker takes over failure of the SA node.

Atrial:	~70bpm. P waves present (may be abnormal, e.g. inverted as originate further down the atria), normal QRS.
Nodal:	~50bpm. Area around AV node takes over, no P wave (no SA node involvement), normal QRS
Ventricular:	~30bpm. QRS broad and has no relationship to P waves

Ectopic extrasystoles may be seen in any rhythm. They *anticipate* expected complexes and come early, unlike escape rhythms that *rescue* and come late. The location of origin of the extrasystole can once again be determined by whether it is broad (ventricular) or narrow (supraventricular). In addition, a supraventricular extrasystole will delay the next P wave, whereas a ventricular extrasystole will not. The easiest to spot are ventricular ectopics which are common:

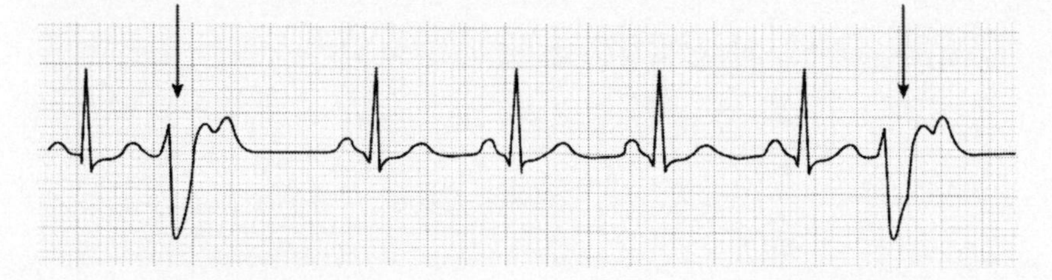

TACHYARRHYTHMIAS

ATRIAL TACHYARRHYTHMIAS

Sinus tachycardia

= Fast rate >100bpm, but normal conduction from SA node

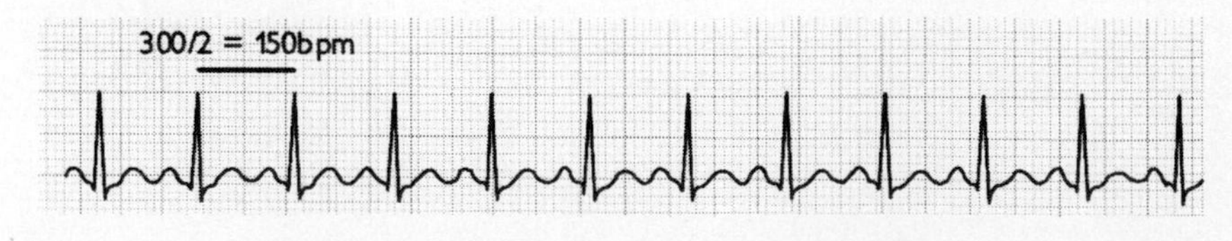

Physiological	↑Sympathetic drive e.g. exercise, fear, pain Stress: infection, hypovolaemia
Pathological	Hyperthyroidism
	Drugs e.g. salbutamol, atropine

<u>**Atrial fibrillation**</u>

= Spontaneous, multiple, rapid, chaotic depolarisation of the atria with conduction to the ventricles limited by the AV node.

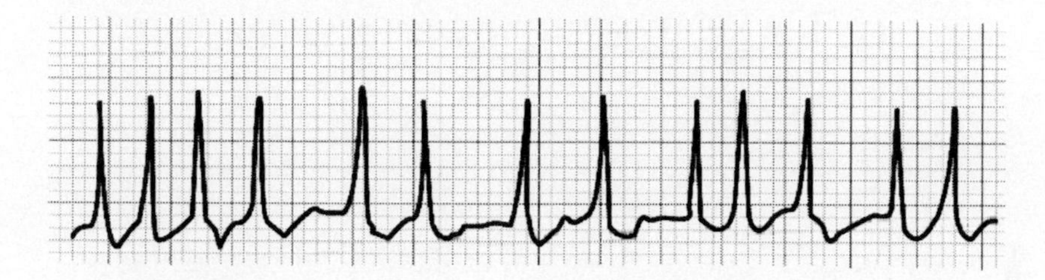

On the ECG, this is seen as:

Irregularly irregular rhythm (chaotic depolarisation)
No P waves (no SA node involvement, erratic baseline)
Narrow QRS complexes (the *origin* is still ATRIAL)

Atrial fibrillation is the one arrhythmia that you should know well.

<u>Aetiology</u>

As there are so many causes of AF, for exam purposes it is better to start with the most common ones:

1. Idiopathic
2. Ischaemic heart disease
3. Hypertension
4. Alcohol
5. Hyperthyroidism

For a more complete list, causes can be divided into:

Cardiac:	*Valves*	Mitral stenosis, mitral regurgitation
	Muscle	IHD, HTN, heart failure, cardiomyopathy
	Pericardium	Pericarditis
Pulmonary:	PE, pneumonia, COPD, cor pulmonale	
Metabolic:	Thyrotoxicosis, alcohol, drugs	

Presentation

EFFECT: *Incidental finding* on ECG, asymptomatic

Symptoms chest pain, palpitations and SOB

Complications stroke (4% risk / year)

CAUSE: *Symptoms* those of underlying disease

Investigations

Simple ECG

Bloods FBC, U&E, cardiac enzymes, TFTs

Imaging Echocardiogram, CXR

Treatment

This can be complex as it depends on the **duration**, **rate** and whether the patient is **haemodynamically compromised**.

In essence:

Treat reversible cause

Rate control

(Preferred in patients over 65 / with coronary artery disease)

Digoxin or β-blockers

Anticoagulate

Patient's risk of stroke determines anticoagulation (NICE 2006)

High risk (should be on warfarin)

>75yrs + HTN/ vascular disease/ diabetes

Previous stroke

Valve disease/ heart failure

Moderate risk (consider warfarin)

>65yrs and no risk factors

HTN/ vascular disease/ diabetes and <75yrs

However risk of bleeding must be taken into account.

Low risk (aspirin alone)

<65 years, no risk factors

Cardiovert -only if AF present <12 months

(Preferred in symptomatic/ heart failure/ younger patients)

Patients must be anticoagulated before cardioversion or have a transoesophageal echo (unless AF is known to have lasted <48 hours) to ensure no clots have formed.

Chemical: Amiodarone

Electrical: Synchronised DC cardioversion -

This is *synchronised* with QRS complex for best results; also, shocking during repolarisation (T wave) can induce ventricular fibrillation.

Atrial flutter

= A re-entry circuit in the right atrium results in rapid, coordinated contraction. Atrial depolarisation occurs at 300bpm and the AV node protectively blocks conduction to the ventricles resulting in 2:1 / 3:1 conduction and a ventricular rate of 150/ 100bpm.

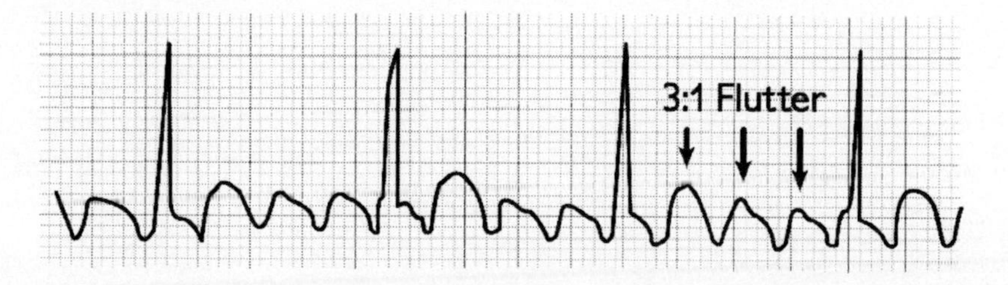

On the ECG this is seen as:

1. Saw-tooth 300bpm P waves
2. Normal, regular QRS complexes (*atrial* in origin) occurring after every 2 or 3 P waves (2:1 or 3:1 conduction)

The aetiology, presentation and treatment is similar to AF, however as there is a focus, radioablative treatment is curative.

JUNCTIONAL TACHYCARDIAS

= Arrhythmias arising from the region of the atrioventricular junction as a result of a re-entry mechanism.

A re-entry mechanism is an abnormal electrical circuit which allows the electrical impulses to cycle around it, thus regenerating the signal so that the SA node is no longer the principle pacemaker.

Unfortunately to confuse the matter, the term "supraventricular tachycardia/ SVT" has been applied to specifically mean junctional tachycardias. Remember, ***any*** tachyarrhythmia arising from the atria or the atrioventricular junction (i.e. above the ventricles) is a supraventricular tachycardia. The most common are atrio-ventricular nodal re-entrant tachycardia (AVNRT) and atrio-ventricular re-entrant tachycardia (AVRT).

AVNRT

This is the most common cause of regular narrow complex tachycardia. It typically affects young patients with no organic heart disease. Its mechanism is relatively complicated; suffice to say that there is a re-entrant circuit around the AV node and bundle of His.

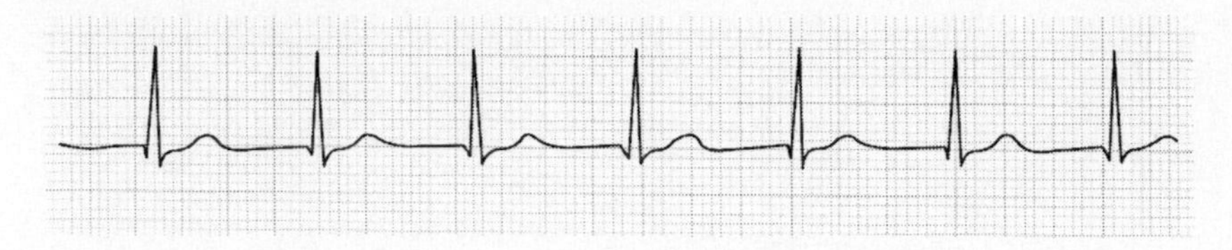

On the ECG this is seen as:

1. Narrow complex tachycardia
2. P waves often absent: they are hidden in the QRS, as atrial and ventricular depolarisation occur at the same time

AVRT

This is where Wolff-Parkinson-White Syndrome comes in. In this case, the re-entry mechanism is due to a congenital addition of tissue which crosses the isolating fibrous ring separating the atria and ventricles. This aberrant connection allows electrical impulses to go back up to the atria and return to the AV node, establishing a circuit. The pathway may also allow atrial depolarisation to bypass the protective AV node. This is dangerous in atrial fibrillation where the ventricular rate depends on the fast atrial rate being attenuated by the AV node.

The resting ECG is typically normal in WPW. If the path allows some of the atrial depolarisation to pass quickly to the ventricle before it gets though the AV node, the ECG may show changes due to early depolarisation of part of the ventricle:

1. Short PR interval
2. Delta wave, a slurred start to the QRS

If also tachycardic, P waves appear absent as they are hidden in the QRS, due to atrial and ventricular depolarisation occurring simultaneously. It may also resemble ventricular tachycardia as the widened QRSs are paired with a fast rate.

VENTRICULAR TACHYARRHYTHMIAS

These rhythms are especially important as they form part of the Advanced Life Support (ALS) pathway for cardiac arrest. Ventricular fibrillation (VF) and ventricular tachycardia (VT) form part of the shockable pathway, while PEA and asystole (not ventricular tachycardias but discussed below) are non-shockable rhythms.

ALS guidelines change continuously as more evidence becomes available. Check with the Resuscitation Council for the latest guidelines.

www.resus.org.uk

Ventricular tachycardia

An abnormally rapid rhythm that originates from the ventricles and thus results in broad, *regular* complexes on the ECG. It is life-threatening because it may result in loss of cardiac output; especially at high rates or in the presence of compromised systolic function. It can also spontaneously degenerate to ventricular fibrillation.

It is most important to ascertain if the patient is cardiovascularly stable and how well the patient is tolerating the VT. They might be happily sitting out eating lunch or in pulseless cardiac arrest. Your response thus varies as appropriate.

Remember pulseless VT is a shockable cardiac arrest rhythm.

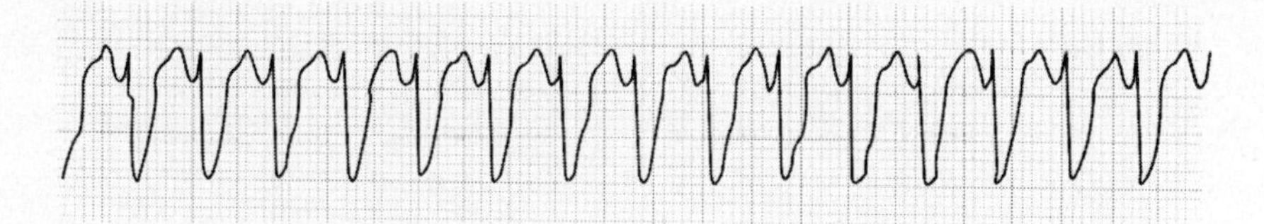

On the ECG VT seen as:

1. Tachycardia >100bpm
2. Broad, *regular* QRS complexes (ventricular in origin)
3. No P waves

Ventricular fibrillation

An abnormally rapid rhythm that originates from uncoordinated, fluttering contractions of the ventricles and thus results in broad, *irregular* complexes on the ECG.

It is life-threatening because the asynchrony of contractions does not allow an adequate cardiac output to be produced (i.e. patient *will* be pulseless).

VF is a shockable cardiac arrest rhythm.

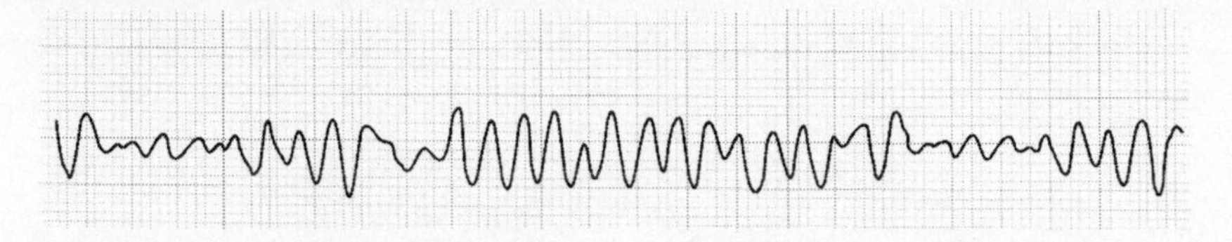

On the ECG VT seen as:

1. Tachycardia >100bpm
2. Broad QRS complexes, *irregular in both time and amplitude*
3. No P waves

Other cardiac arrest rhythms

These rhythms are on the ***non-shockable*** branch of the ALS protocol.

Asystole

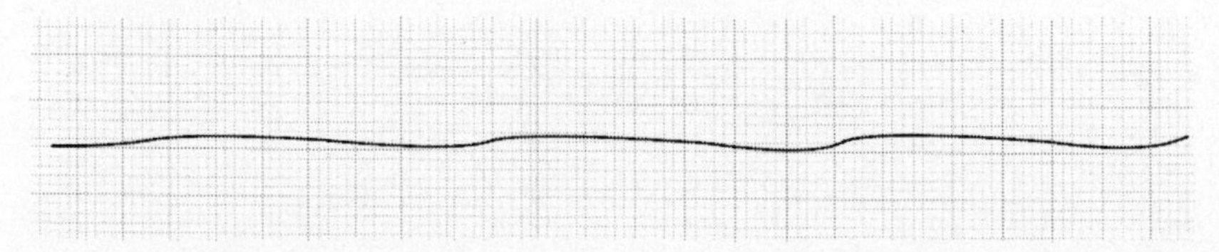

No complexes, baseline drift. If the ECG line is completely straight (like the line they show when someone dies on a TV show) check the leads which will most likely be disconnected.

Pulseless electrical activity

Electrical activity (any except VF/VT) but no cardiac output, determined by feeling for a central pulse. NB the ECG may look entirely normal!

SICK SINUS SYNDROME

These are episodes of bradycardia and tachycardia, therefore both are treated, the bradycardia with pacing, and the tachycardia with rate control medication.

Hypertension

Definition

A major risk factor for stroke and ischaemic heart disease defined as a blood pressure over 140/90mmHg on two separate occasions. However, the risk is continuously related to blood pressure.

Description

Incidence Extremely common

Age Increases with age

Sex M>F

Geography Worldwide, increased risk in Blacks and Asians.

Aetiology

Primary hypertension (90%) = "Essential hypertension"

This has a multifactorial aetiology

Genetic Family history

Foetal Low birth weight

Environmental Obesity, in particular ↑waist:hip ratio

Alcohol >6units/day

↑ Sodium intake

Stress

Insulin resistance / metabolic syndrome / syndrome X

Insulinaemia

Decreased glucose tolerance

↑HDL & TG

Central obesity

Secondary hypertension

Renal Diabetic nephropathy

Renovascular disease

Polycystic kidney disease

Chronic glomerulonephritis

Endocrine	Cushing's syndrome
	Conn's syndrome
	Phaeochromocytoma
	Acromegaly
	Congenital adrenal hyperplasia
Drugs	Oral contraceptive pill
	Steroids
	MAO inhibitors (cheese reaction)
CVS	Coarctation of the aorta
Pregnancy	Pre-eclampsia

Pathophysiology

The pathogenesis of hypertension is unclear. In chronic hypertension, cardiac output is normal while total peripheral resistance is raised. This leads to:

1. Structural vessel wall changes:

   ↑ Wall thickness, ↓ lumen diameter

2. Mechanical stress promoting atherosclerosis
3. Left ventricle hypertrophy due to ↑ afterload
4. Decreased renal perfusion leads to ↓GFR, renin release and subsequent vasoconstriction and water retention, further increasing blood pressure

History and Examination

EFFECT: Asymptomatic incidental finding or complications

CAUSE: Risk factors as above for primary/secondary hypertension

Complications relate to end-organ damage

Brain	Stroke (thrombo-embolic and haemorrhagic)
Heart	IHD, heart failure
PVS	Stenosis and aneurysms
Kidneys	Renal failure
Retina	Blindness

The features of hypertension on fundoscopy are:

Grade I	*Silver-wiring* =tortuous arteries	
Grade II	Above	+ *AV nipping* = thick arteries overlying veins
Grade III	Above	+ *Flame haemorrhages*
		+ *Cotton wool spots/soft exudates* =small infarcts
Grade IV	Above	+ *Papilloedema* = malignant hypertension

Investigations

Ambulatory blood pressure monitoring if necessary to confirm diagnosis.

Further tests aim to identify:

1. Secondary causes
2. End-organ damage
3. Concomitant conditions (diabetes!) that further increase the risk of vascular disease

Simple	ECG	LVH
	Urinalysis	Renal disease, Phaeochromocytoma Cushing's
	BM	Diabetes
Bloods	U&E	Renal and endocrine disease
	Cholesterol	
	Glucose	
Imaging	CXR	Heart failure, Coarctation of aorta
	Echocardiogram	Heart failure
	Angiography	Renal artery stenosis
	MRI	Coarctation of aorta

Treatment

The aims are to reduce blood pressure as well as other cardiovascular risk factors.

Conservative treatment

Reduce blood pressure

Exercise and weight loss

Reduce heavy alcohol intake

Restrict salt intake

Reduce other cardiovascular risk factors

Smoking cessation

Healthy diet avoiding high cholesterol foods

Control diabetes

Medical treatment

The Cambridge **A(B)/CD** rule

A	ACE inhibitors/ Angiotensin II receptor antagonists
(**B**	β-blockers)
C	Calcium channel blockers
D	Diuretics

A works best for younger patients (<55), while **C** or **D** are preferred in the elderly or Blacks. Monotherapy or combinations of ABCD may also be used as necessary, whilst avoiding the combined use of β-blockers and diuretics (diabetogenic). Most patients will require two or more drugs to control their hypertension.

Note that β-blockers have fallen out of favour as they appear to offer no advantage for preventing MI or death and are no longer recommended as first-line treatment of primary hypertension.

	Younger patients/ non-Blacks	Older patients/ Blacks
Step 1	**A**	**C or D**
Step 2	**add C or D**	**add A (or B)**
Step 3	**A+C+D**	**A+C+D**

Spironolactone and α-blockers may be useful in resistant hypertension

ACE inhibitors are favoured in diabetic and heart failure patients, while β-blockers are used with caution in heart failure. Angiotensin II receptor antagonists are useful in those patients who develop cough with ACE inhibitors.

Prognosis

Prognosis is variable and depends on age at presentation, cause and level of blood pressure, response to treatment, end-organ damage and co-existing risk factors.

British Hypertension Society 2004 Guidelines

Primary prevention:

<140/90	Lifestyle advice and reassess at 5 years
>140/90	
No IHD/stroke risk	Lifestyle advice and reassess at 1 year
IHD/stroke risk	Lifestyle advice and medical treatment
>160/100	Lifestyle advice and medical treatment

Secondary prevention:

Everyone who has had a cardiovascular event should be treated

Target blood pressure is < 140/85 but depends on concomitant disease.

Diabetics: <130/80

Proteinuria: <125/75

Blood pressure must be reduced slowly to avoid complications.

MALIGNANT HYPERTENSION

Definition: Severe high blood pressure of rapid onset with a diastolic pressure >140mmHg

Aetiology: Secondary causes of HTN, in particular phaeochromocytoma, Conn's Syndrome, renal artery stenosis and pre-eclampsia

Pathophysiology: Fibrinoid necrosis of blood vessel walls leading to

Renal failure

Cerebral oedema and haemorrhage resulting in encephalopathy

Retinal changes (papilloedema)

History and Examination: Severe headache, visual disturbances, seizures, loss of consciousness, cardiac failure and signs and symptoms of underlying cause.

Treatment: Avoid sudden drops in blood pressure as this may lead to end-organ infarction, blood pressure should be reduced over *days*. The treatment of the hypertension itself is medical if no encephalopathy and intravenously with central line monitoring if hypertensive encephalopathy is present.

Treatment depends on the cause:

Phaeochromocytoma: surgical excision, α-blockers. Never β-blockers alone.

Conn's Syndrome: surgical excision, spironolactone

Renal artery stenosis: avoid ACEi if bilateral, cautious use if unilateral

Prognosis: Untreated malignant hypertension has a 20% one year mortality.

HEART FAILURE

Definition

The inability of the heart to maintain sufficient cardiac output to meet demand, defined as an ejection fraction of less than 45%.

Description

Incidence	Common -10% of over 65 year olds
Age	↑ with age

Aetiology

Heart failure is the final common pathway of all heart disease. Indeed, it can be caused by any disease that increases *myocardial work* hence it cannot be written under 'cause of death' on a death certificate since it is a mode, rather than a distinct cause.

↑ Preload	*Volume overload*	Mitral/aortic regurgitation
		Fluid overload
Pump failure		
Muscle	*Myocardial dysfunction*	Ischaemic heart disease
Electricity	*Altered rhythm*	Arrhythmias, -ive inotropes
Volume	*Compromised filling*	Pericarditis, cardiac tamponade

↑ **Afterload**	*Obstruction to outflow*	Aortic stenosis, Coarctation of aorta
	Obligatory high output	Hypertension, anaemia, thyrotoxicosis

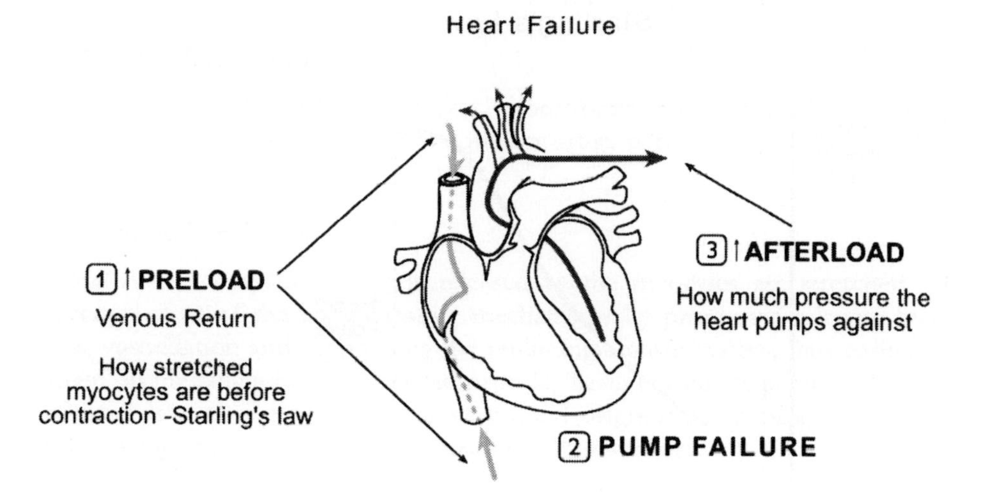

Pathophysiology

Initially, the heart compensates through physiological mechanisms. Eventually, these same protective mechanisms spiral out of control and decompensate by increasing the work of a struggling heart to maintain cardiac output so heart failure ensues. As the cardiovascular system continues its attempts to compensate, the heart is pushed further into failure.

1. Increase in preload

As preload (end diastolic volume) increases, so does contractility according to Starling's curve, however, a point is reached where the physiology decompensates (down-sloping part of the curve) and at high preloads, the ejection fraction decreases. Cardiac output is mostly maintained by increasing total peripheral resistance and heart rate, thus contributing to further heart failure.

Starling's law can be demonstrated by the following, simple experiment.

Imagine squeezing a golf ball in the palm of your hand. It is difficult to exert maximal force due to the small volume. Now imagine squeezing a tennis ball in the palm of your hand. Its greater volume stretches your muscles allowing you to exert a larger force.

1. Right heart failure

EFFECT: *Signs and symptoms of right heart failure*

Right heart failure is a disease of *signs.* Although symptoms are vague and may include fatigue, shortness of breath and anorexia, signs of fluid congestion are plentiful:

↑ JVP

Hepatomegaly (smooth, pulsating)

Pitting oedema

Ascites

Right-sided pleural effusion (transudate)

CAUSE: *Signs and symptoms of underlying cause*

↑ *Preload*	Right-sided valve disease
	Left to right shunts increase pressures in the right heart
Pump failure	Ischaemic heart disease, cardiomyopathy, arrhythmias
↑ *Afterload*	Left heart failure (mcc of RHF!)
	Pulmonary hypertension
	Mitral stenosis resulting in pulmonary HTN

2. Left heart failure

EFFECT: *Signs and symptoms of left heart failure*

Left heart failure is a disease of symptoms. You should ask about all of the following symptoms of pulmonary oedema for every cardiac history.

<u>Dyspnoea</u>

New York Heart Association Class

I	Asymptomatic during ordinary activities
II	Symptoms during ordinary activities
III	Symptoms during less than ordinary activities
IV	Symptoms at rest

Orthopnoea

This is the inability to lie flat without getting short of breath, best assessed by asking how many pillows a patient sleeps with at night, e.g. "three pillow orthopnoea".

Paroxysmal nocturnal dyspnoea

These are, translated literally, sudden episodes of night-time breathlessness, best assessed by asking the patient "do you ever wake up breathless at night?"

Signs of left heart failure are seen late in disease and include:

Pulmonary oedema	Bibasal lung crepitations
Cardiomegaly	Displaced apex beat
Gallop rhythm	3rd/4th heart sound with a tachycardia
3rd HS	Volume overloaded ventricle
	"Ken-tuc-ky"
4th HS	Atrial contraction against stiff ventricle
	"Ten-nes-see"

CAUSE: *Signs and symptoms of underlying cause*

↑ *Preload*	Left-sided valve disease (not mitral stenosis -left *atrium* fails)
Pump failure	Ischaemic heart disease (mcc), cardiomyopathy, arrhythmias
↑ *Afterload*	Hypertension, coarctation of the aorta

3. Congestive heart failure

This is a term used to describe right heart failure secondary to left heart failure and therefore carries both sets of signs and symptoms. It is also known as "biventricular heart failure".

4. Cor Pulmonale

This is a term used to describe right heart failure secondary to pulmonary *hypertension* this means that it is secondary to not only *lung* disease but *any* disease that causes a downstream circulation problem. The precise mechanism therefore varies. The most common causes are pulmonary embolism acutely and COPD chronically.

EFFECT: *Signs and symptoms of pulmonary hypertension and right heart failure*

Hx: Chest pain, SOBOE, syncope, sudden death

OE: Signs of pulmonary hypertension

↑ JVP with prominent a and v waves

Loud pulmonary 2nd heart sound

Pulmonary regurgitation (Graham Steell murmur)

Signs of right heart failure

CAUSE: *Signs and symptoms of underlying cause of pulmonary hypertension*

Lung vasculature	Pulmonary embolus
	Primary pulmonary hypertension
	Pulmonary artery stenosis
Chronic under ventilation	
Lung parenchyma	COPD and chronic lung diseases
Musculoskeletal	Kyphoscoliosis, polio, myasthenia gravis, obstructive sleep apnoea
↓ Respiratory drive	CVA, drugs
Left heart disease	Left heart failure
	Mitral stenosis
	Left atrial myxoma

5. Acute heart failure

EFFECT: *Acute signs and symptoms of heart failure*

As this is an acute disease, signs that take a long time to develop such as hepatomegaly and ankle oedema are not seen. The pulmonary oedema that develops due to <u>left</u> heart failure is a ***medical emergency*** and may lead to cardiac arrest.

Symptoms include SOB, orthopnoea, and pink, frothy sputum

Signs include ↑ JVP, bibasal crepitations, wheeze ("cardiac asthma") and a sympathetic response (pale, sweaty, tachycardic) in a patient that is sitting up, leaning forward and agitated -the patient is technically drowning in his own fluid!

CAUSE: *Signs and symptoms of underlying cause of acute heart failure*

↑ *Preload*	Fluid overload
Pump failure	Myocardial infarction (mcc)
	Papillary / chordal rupture
	Cardiac tamponade
	Arrhythmias
↑ Afterload	Pulmonary embolus

6. High output heart failure

The definition of heart failure is failure of the heart to produce an adequate cardiac output *to meet demand.* In high output heart failure, which is rare, the heart is unable to maintain the increased cardiac output that disease demands.

Causes include anaemia, thyrotoxicosis and septicaemia.

<u>Investigations</u>

The aims of investigation are to determine the cause and exacerbating factors of heart failure as well as to assess the severity.

Simple	**ECG**	Ventricular hypertrophy
		Ischaemia, arrhythmias
	BM	Diabetes
	Urine dipstick	Diabetes
Blood	FBC	Anaemia / sepsis
	U&E	Fluid overload, renal failure
	Glucose	
	Cardiac enzymes	
	Thyroid function tests	
	BNP	70% sensitive but not specific for LVF
		If normal, HF much less likely

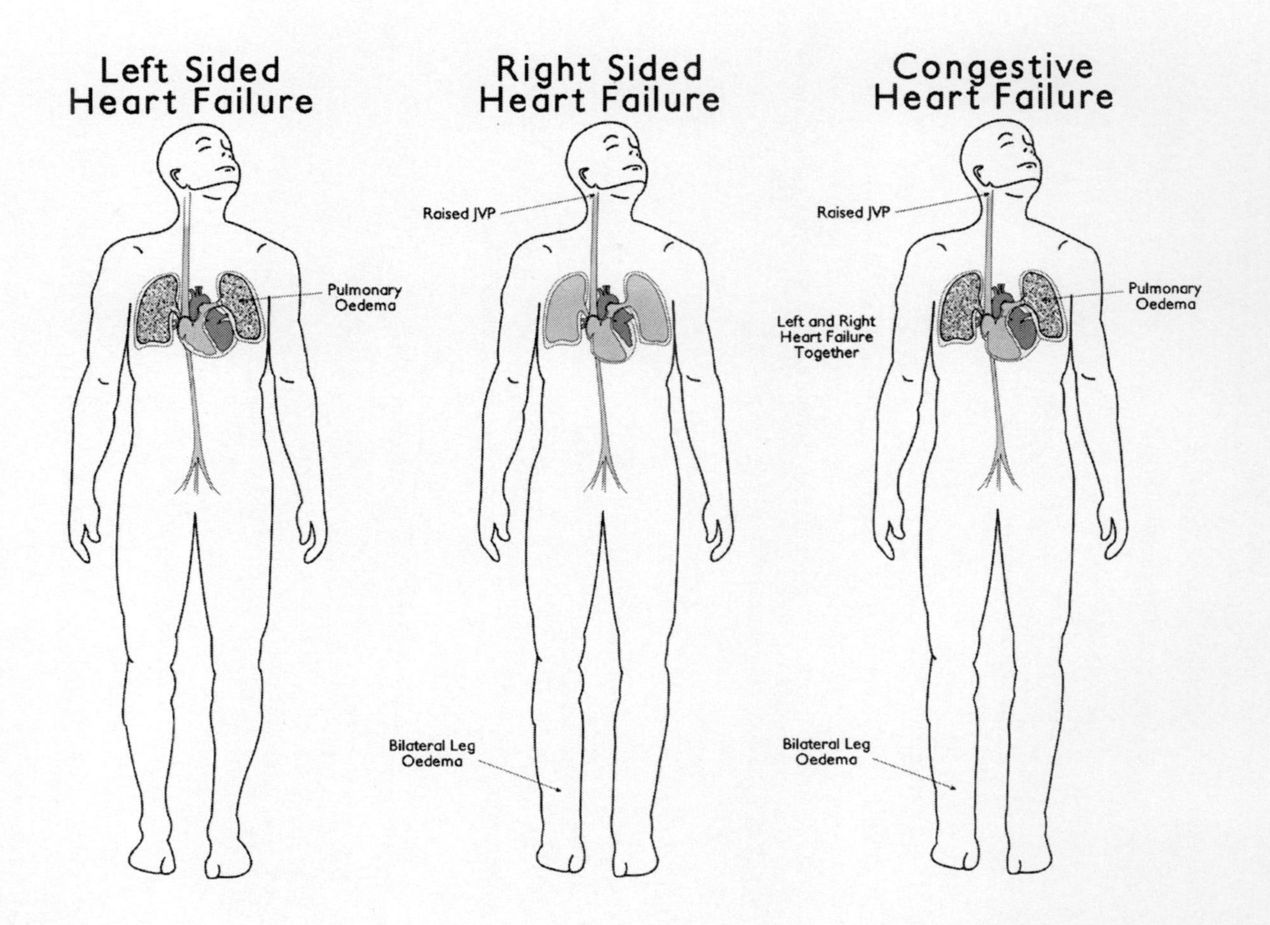

Left Sided Heart Failure
Pulmonary Oedema
Right Sided Heart Failure
Raised JVP
Bilateral Leg Oedema
Congestive Heart Failure
Raised JVP
Pulmonary Oedema
Left and Right Heart Failure Together
Bilateral Leg Oedema

Imaging	**CXR**	"ABCDEF"
		Perihilar **A**lveolar oedema "bat's wings"
		Kerley **B** lines (interstitial oedema)
		Cardiomegaly (>50% cardiothoracic ratio)
		Upper lobe blood **D**iversion
		Pleural **E**ffusion
		Fluid in the **F**issures
	Echocardiogram	*gold standard
		Defines structural abnormalities
		Quantifies function (ejection fraction)

If **BNP levels or ECG** are abnormal, then an **echocardiogram** should follow (NICE guidelines). If these, along with a CXR are normal, then heart failure is unlikely.

Treatment

1. Acute heart failure

This is a medical emergency

Airway Check and secure with adjunct if necessary

Breathing Oxygen: 15L through non-rebreather mask

Sit patient up

Circulation Gain i.v. access

Attach cardiac monitor

Treat effect

Diamorphine 2.5-5mg i.v. + anti-emetic (sedates + vasodilation)

Frusemide 40-80mg i.v. (vasodilation + diuretic)

GTN sublingual or infusion, caution if low BP (venodilation)

CPAP

Treat cause

Identify and treat underlying cause/ exacerbating factors:

e.g. Arrhythmias/MI

Continue investigations once patient is stable, particularly a chest X-ray to confirm the diagnosis and later an echocardiogram to assess left ventricular function.

2. Chronic heart failure

The aims of treatment are to:

Treat the underlying cause

Relieve symptoms

Slow progression

Improve survival

Conservative (also preventative)

Smoking cessation

Physical activity

Endurance exercise i.e. 30mins walk 3-5 times a week
Short periods of bed rest for exacerbations (avoid DVT)

Diet

Salt restriction
Fluid restriction in severe heart failure
Weight reduction
Avoid alcohol (-ive inotrope)

Medical

Proven prognostic benefit, ABC

<u>A</u>CE inhibitors/ Angiotensin II receptor antagonists

Improve symptoms and prolong life
ATII antagonists useful if develop cough with ACEi (1:10)
Beware "first dose hypotension" with ACEi: take before bed

<u>B</u>eta-blockers

Improve symptoms and prolong life
Start low, go slow as may decrease ejection fraction initially
Do not give in acute heart failure

Spironola<u>C</u>tone (K-sparing diuretic)

Reduces mortality by 30% when added to above treatment
Some symptomatic relief
Beware hyperkalaemia (small risk)

Nitrate (e.g. isosorbide mononitrate) plus hydralazine

Second line, useful if contraindication to ACEi

Mortality decreased only if used together

Symptomatic relief, no effect on prognosis

Frusemide (Loop diuretics)

Routinely used to provide symptom relief, *doesn't* alter outcome

Beware hypokalaemia

Digoxin

Second line (even if in sinus rhythm) for symptom control

In summary, patients with clinical heart failure can be on an ACEi/ATII antagonist, a β-blocker (once stable) and spironolactone to improve prognosis and a loop diuretic (frusemide) for symptomatic relief.

Negative inotropes (such as calcium channel blockers) should be avoided.

Exacerbating factors should be treated: Arrhythmias, anaemia, thyrotoxicosis

Cardiovascular risk factors should be addressed: HTN, ↑ cholesterol, diabetes.

Interventional

Cardiac transplantation

Artificial hearts (used as bridge to transplant)

Left ventricular assist devices (used as bridge to transplant)

Prognosis

The prognosis of heart failure is generally poor, with 50% mortality at 5 years.

Ischaemic heart disease

Definition

Ischaemic heart disease is the insufficient supply of oxygen to the myocardium.

IHD is a progressive disease. The first symptom is often stable angina which may progress to an acute coronary syndrome.

Description

Incidence	Extremely common
Age	Incidence and prevalence ↑ with age
Sex	M>F until menopause, after this a woman's risk increases to equal that of a man (M=F)
Geography	The leading cause of death in the western world

Aetiology

Decreased oxygen supply

- **↓ Blood flow**
  - Hypotension
  - ***Obstruction*** (mcc)
    - Thrombosis (**atherosclerosis**)
    - Embolism
    - Spasm (Prinzmetal's)
    - Arteritis (SLE)
- **↓ O_2 content**
  - Hypoxia/ respiratory disease
  - Anaemia
  - Carboxyhaemoglobulinaemia

Increased oxygen demand

- Thyrotoxicosis
- Myocardial hypertrophy (e.g. hypertension, aortic stenosis)

Risk Factors

Modifiable risk factors

1. Hypertension
2. Smoking
3. ↑ Cholesterol
4. Diabetes
5. Obesity

Fixed risk factors include: family history, age, sex (M>premenopausal F)

Pathophysiology

Atherosclerosis

Atherosclerosis is a complicated *inflammatory* process involving the accumulation of *lipid, macrophages and smooth muscle cells* in the *intima* of medium and large arteries. Subsequent rupture of an *unstable plaque*, can result in *thrombus* formation, which can itself break off leading to an *embolus*. The leading hypothesis for the cause of atherosclerosis is the 'response to injury' hypothesis, which works progressively as follows:

1. **Endothelial injury**

   Mechanical (particularly at bifurcations, ↑ turbulent flow)
   Cholesterol
   Immunological

2. **Lipoprotein uptake by macrophages**

   Foam cells, contained by fibrous cap

3. **Smooth muscle migration and proliferation**

4. **Plaque rupture**

   Platelet adhesion leads to thrombus formation ± embolisation

This can occur in any medium / large artery, so if there is vascular disease in one place, you should think about there being disease elsewhere. Regions affected include:

Heart	Coronary artery disease (angina, ACS)
Brain	Cerebrovascular disease (ischaemic stroke, TIA)
Kidneys	Renal artery disease (renal failure, hypertension)
Peripheral	Peripheral artery disease (thrombo-embolism, aneurysm)
Gut	Ischaemic bowel (splenic flexure has poor blood supply)

History and Examination

EFFECT:

Angina

Angina is the *symptom* of central crushing chest pain which may radiate to the jaw or, typically, left arm and represents myocardial ischaemia.

Exertional/ Classical angina (mc)

This type of angina occurs on exertion and is relieved by rest. It reflects inadequate oxygenation of the myocardium due to an increased oxygen demand which is unable to be met due to coronary artery disease. It can be graded using the *Canadian Cardiovascular Society scale:*

Grade I	Symptoms on strenuous activity
Grade II	Symptoms on moderate activity (stairs)
Grade III	Symptoms on mild activity (flat)
Grade IV	Symptoms at rest (unstable angina)

Stable angina does <u>not</u> form part of Acute Coronary Syndromes

You must differentiate between stable angina and acute coronary syndromes as this greatly influences management and prognosis.

ACS (discussed below) includes:

Unstable angina

Non-ST elevation MI (NSTEMI)

ST elevation MI (STEMI)

Decubitus angina

This type of angina occurs on lying flat and is associated with decreased left ventricle function due to coronary artery disease.

Variant/ Prinzmetal's angina

This occurs at rest and is typically seen in women. It is due to spasm of the coronary arteries and may lead to arrhythmias, (ventricular tachycardias and heart block).

CAUSE: Evidence of risk factors may be elicited in the history or seen on examination. Don't forget to mention risk factors in your examination such as:

High BMI	Obesity, especially centripetal
Tar stained fingers	Smokers
Lipid arcus	Blue ring around iris (may be normal in elderly)
Xanthelasma	Waxy cholesterol plaques around eyes
Xanthoma	Cholesterol deposits on extensor tendons (Achilles)

Remember to exclude *aortic stenosis* as a cause of angina.

Investigations

The aim of investigation is to diagnose and determine the extent of ischaemic heart disease as well as quantify cardiac function. Biochemical risk factors are also assessed. This allows risk of myocardial infarct to be gauged and guides treatment.

Simple	**ECG**	*Resting*	Old MI, LBBB
		Exercise	Ischaemia suggested if: ST depression >1mm + chest pain But ↑ *rate of false +ives and -ives*
Blood	Glucose		
	Lipids		
Imaging:			

Echocardiography

Assesses structure and cardiac function

A stress echo with dobutamine can detect ischaemia

Perfusion imaging with thallium assesses regional flow

Angiography:

Allows planning for revascularisation (CABG/PTCA)

Visualises coronary anatomy

Identifies areas of stenosis

Haemodynamically significant according to:

Length of vessel affected

Radius of narrowing

If areas of short stenosis in accessible vessels: **PTCA**

If multiple, long areas of severe stenosis and difficult anatomy: **CABG**

Angiography has a mortality of <1/1000

Essentially, patients should go for exercise stress testing, or alternatively a stress echo if unable to exercise on the treadmill. A strongly positive result should then be investigated with angiography with a view to offer invasive revascularisation.

Treatment

The aims of treatment are to:

- Reduce risk factors
- Relieve symptoms
- Improve survival

Conservative

Weight loss and exercise

Low cholesterol diet

Smoking cessation

Medical

Treat hypertension and control diabetes

Proven prognostic benefit

Aspirin reduces risk of coronary events

Lipid lowering drugs

Statin	↓ LDL, ↑ HDL
Fibrate	↓ Triglycerides

Target cholesterol –JBS2 (Joint British Society's Guidelines, 2005)

Total cholesterol	<4 mmol/L
LDL	<2 mmol/L
HDL	>1 mmol/L

<u>*Symptomatic relief by reduction of myocardial O_2 demand*</u>

Immediate relief

Glyceryl trinitrate (GTN)	↓ Venous return and preload
Preventors	
β-blockers	↓ Heart rate and contractility
Long acting nitrates	↓ Venous return and preload
Calcium channel blockers	↓ Contractility and vasodilate: ↓ TPR, ↓ afterload

Interventional

Percutaneous transluminal coronary angioplasty (PTCA) ± stenting

Angioplasty = stenosed artery dilated using balloon

Stenting = stent deployed to maintain artery open

Provides good symptom control, no reduction in mortality

Complications

Death (1%)

Acute myocardial infarction (2%)

Conversion to CABG (2%)

Local complications at guidewire insertion site

Restenosis (~25% within 6 months)

Note: i.v. glycoprotein IIb/IIIa inhibitors during procedure can ↓ ischaemic complications while aspirin and clopidogrel ↓ stent thrombosis. PTCA should be carried out where facilities exist for conversion to emergency CABG.

Coronary artery bypass graft (CABG)

Indications

Symptomatic relief only

Symptoms uncontrolled by medical treatment
\+ Unsuitable for PTCA

Improved survival and symptomatic relief

Left main stem disease
Three vessel disease including proximal LAD
Occlusion of coronary artery following angioplasty

Graft

Artery	Internal mammary artery
Vein	Reverse saphenous vein
	(Reversed because veins have valves!)

Complications (specific)

Immediate	Death (<1%)
Early	Myocardial infarction
	Pleural effusion
	Post-CABG syndrome
	CVA
Late	Re-stenosis
	Especially vein grafts >50% in 10 years
	-delayed by low dose aspirin

Prognosis

The prognosis is good with the annual mortality of angina being less than 2%

ACUTE CORONARY SYNDROMES

Definition

ACS describes a spectrum of myocardial ischaemia. It is the continuum between unstable angina and complete ST elevation myocardial infarction (STEMI).

ACS is split into three groups according to ECG changes and troponin results:

Unstable angina

Non-ST elevation MI **(NSTEMI)**

ST elevation MI **(STEMI)**

To diagnose ACS, you need two out of three components:

1. Cardiac chest pain
2. ECG changes
3. Cardiac enzyme rise

Aetiology & Pathophysiology

ACS is an endpoint of ischaemic heart disease. The aetiology and pathophysiology are as outlined above.

Diagnosis

1. History

Try to characterise the pain as much as possible -does it sound cardiac?

"Central crushing chest pain radiating to the jaw/ arm"

Usually severe, but there may be no pain at all

(Diabetics, elderly)

Exertional

Does it occur on exertion and improve with rest as oppose to pleuritic pain which is worse on breathing in or musculoskeletal pain which is tender and occurs with specific movements or heartburn which is worse on lying flat and after meals?

Unstable angina:

Symptoms occurring at rest

Symptoms worsening over < 1 month (crescendo angina)

Evokes a sympathetic response

Any nausea, vomiting, sweating, clamminess?

Response to GTN

Often patients will have received GTN in the ambulance. Taking the clinical scenario into account, no response at all in a patient who has a normal ECG and says the pain is worse on lying flat is a good indicator that the pain is not cardiac. If the patient is in severe pain, has ST elevation and has not responded to GTN, this indicates that they need further intervention fast.

Timing

You need to know the exact time of onset as this influences potential for thrombolysis/ time troponin is checked

2. ECG changes

ECG changes tell you if there is ischaemia. Remember that Q waves (indicating full-thickness MI) take time to develop, so "Q wave" or "non Q wave" MI is a diagnosis given on discharge.

Initially we focus on "ST elevation" or "non ST elevation" ECG changes to stratify each patient's risk as the results of blood tests for troponin levels (which should be done 12 hours after the pain started) aren't known and Q waves haven't had time to develop. Those with ST elevation are bound to have positive troponin, indicating muscle damage. If however, we are able to treat the STEMI in time and save myocardium, the troponin may come back negative. This is termed an "aborted MI".

Transient changes

ST depression/ T wave inversion	Ischaemia Unlikely to progress to Q waves
ST elevation	Ischaemia/ infarction Likely to progress to Q waves

Permanent changes

Q waves	Full-thickness MI

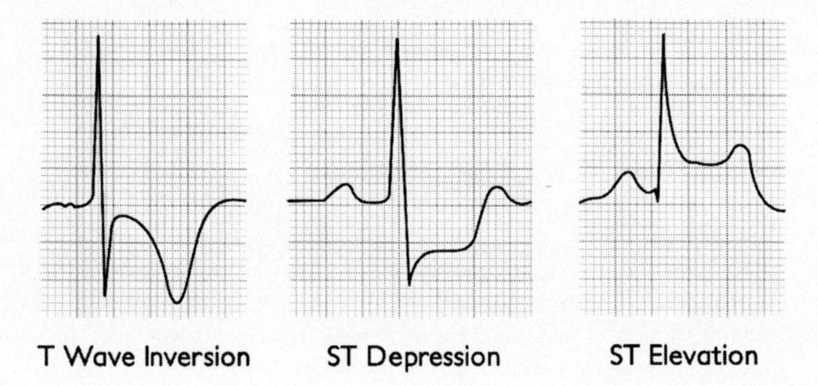

The location of the changes tells you which part of the heart is affected as shown in the diagrams below. These depict:

1. The position of the chest leads and the parts of the heart they investigate
2. The position of the limb leads and the parts of the heart they investigate
3. The anatomy of the coronary arteries

Putting all of these together will allow you to pinpoint the location of the lesion.

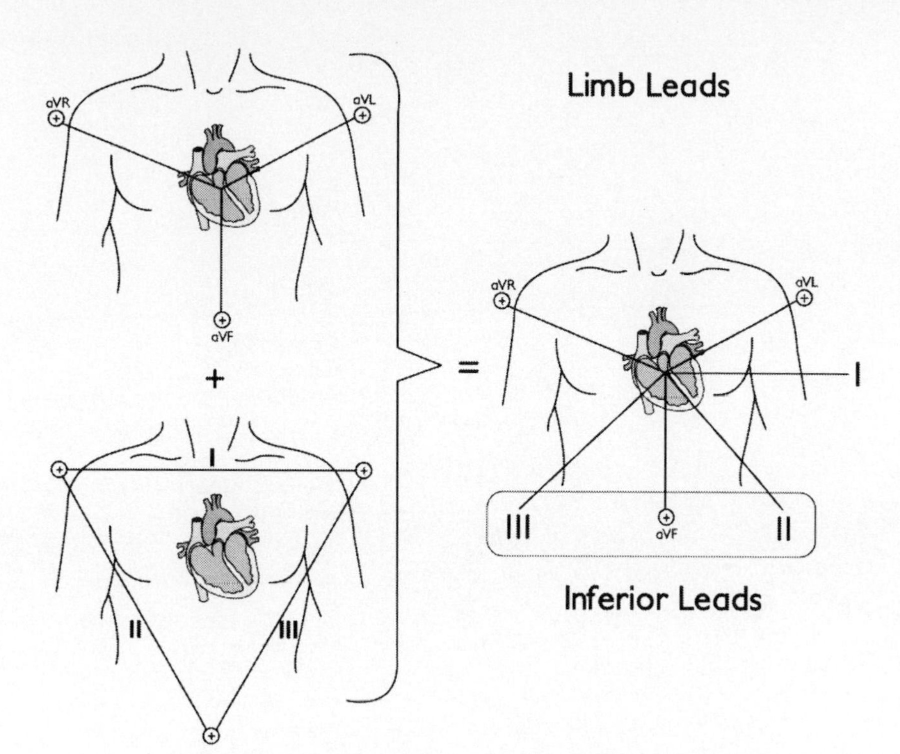
Limb Leads
aVR
aVL
aVF
+
I
II
III
=
aVR
aVL
I
III
aVF
II
Inferior Leads

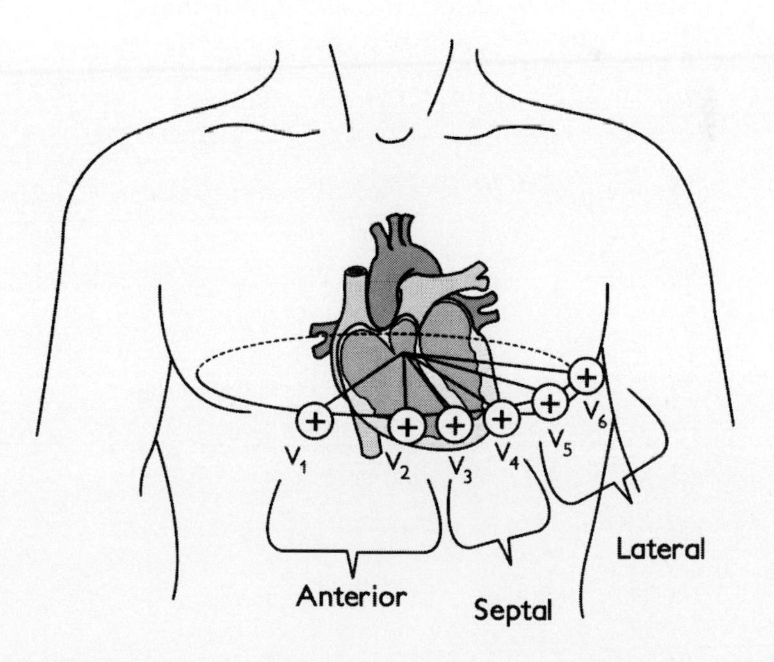
Chest Leads
V1
V2
V3
V4
V5
V6
Anterior
Septal
Lateral

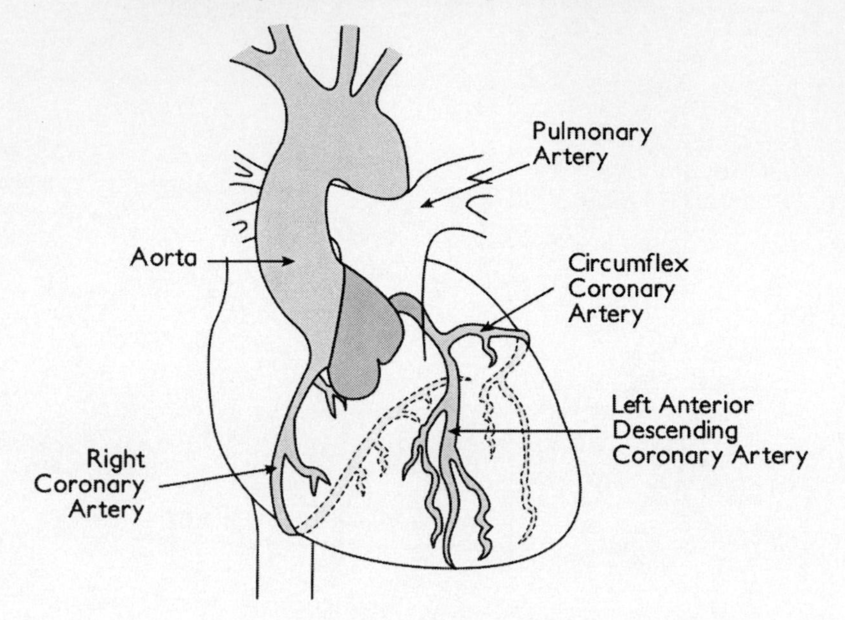

Leads	Area	Supplied by	
V1-2	Anterior	LCA:	Diagonal branch of LAD
V3-4	Septal	LCA:	Septal branch of LAD
V5-6	Lateral	LCA:	Left circumflex artery
V1-6	Anterolateral	LCA:	Left main stem disease
II, III, aVF	Inferior	RCA:	Posterior descending branch
V1-3	Posterior*	LCA:	Circumflex
		RCA:	Posterior descending branch

LCA= left coronary artery, RCA= right coronary artery

*As there are no posterior leads, you see reciprocal changes in V1-3:

-Positive R wave in V1

-ST depression V1-3

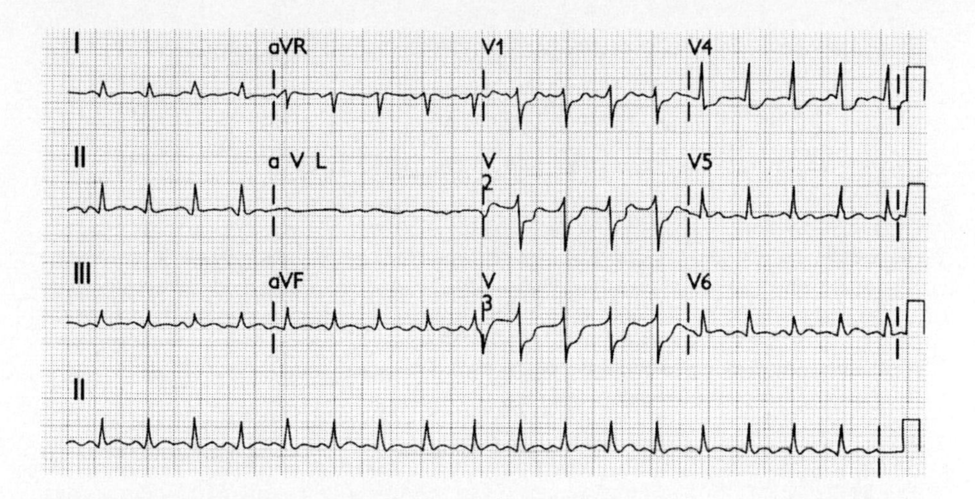

This ECG shows ST depression in leads V1-V4. This indicates *ischaemia* in the area supplied by the left anterior descending artery. The lateral leads (V5-V6) and inferior leads (II, III & aVF) are unaffected. A positive troponin in this patient would confirm an anteroseptal *infarct.*

3. Cardiac enzymes

These are released by necrotic myocytes and therefore indicate that heart muscle has been damaged i.e. infarction and not just ischaemia.

Troponins T & I

Sensitive and specific to myocardium

Released within 2-4 hours, persist for days

Best measured 12 hours after pain starts

Most hospitals use this as the marker for cardiac damage

Creatine kinase (CK)

Non-specific

CK-MB cardiac isoform can be measured

Peaks within 24 hours, normal by 48 hours

Falling out of favour

<u>**Summary**</u>

The exact definitions remain hazy as this is a spectrum. The patient is initially treated as an emergency for "ACS" and according to risk. Full diagnoses are given later during the admission once more information is at hand.

Unstable angina

Good cardiac history
±ST/T wave changes but no ST elevation
ECG may even be normal
Negative Troponin

=ischaemia but no myocardial damage

NSTEMI

ST/ T wave changes but no ST elevation
Positive Troponin

=ischaemia + infarction
Unlikely to be full thickness
(don't develop Q waves)

STEMI

ST elevation
Positive Troponin

=ischaemia + infarction
Likely to be full thickness (develop Q waves)

		ECG: ST elevation	
		+	-
Troponin	+	STEMI	NSTEMI
	-	Aborted MI	Unstable angina

Investigations

Simple **ECG** ***Resting*** As above
Exercise For haemodynamically stable, low risk cases (troponin –ive)

Blood Glucose, Lipids — Cardiac risk factors

Imaging

Echocardiography

Assesses structure and cardiac function
A stress echo with dobutamine can detect ischaemia
Perfusion imaging with thallium to assess regional flow

Angiography

Visualises coronary anatomy, identifies areas of stenosis

Treatment

Initial assessment should be quick allowing immediate initiation of treatment to relieve symptoms, limit cardiac damage and lower the risk of cardiac arrest.

Initial treatment for all ACS is "MONA"

Morphine	-with anti-emetic
Oxygen	-high flow via non-rebreather
Nitrates	-GTN spray/ tablet. Infusion if ongoing pain
Aspirin	-300mg stat

Early β-blockade has also been shown to improve symptoms and reduce ischaemia.

1. STEMIs and reperfusion therapy

The aim of reperfusion therapy is to restore blood supply to the cardiac muscle that has not yet been irreversibly damaged.

Thrombolysis

Due to risks of haemorrhage, strict indications and contraindications to thrombolysis exist. It is most likely to be beneficial in large, anterior infarcts presenting within the first few hours. Successful thrombolysis is defined by the cessation of pain and resolution of ST elevation by more than 50%. Door to needle time is an important factor, so much so that emergency departments started talking about decreasing "*handbrake* to needle time". Now pre-hospital thrombolysis is being initiated. "Aborted MIs" is the term used where patients with STEMIs who are thrombolysed immediately go on to have a *negative* troponin, indicating no damage to the myocardium.

Indications

<12 hours chest pain
ST elevation >2mm in 2 adjacent chest leads (V1-6)
>1mm in 2 adjacent limb leads (II, III, aVF)
New onset left bundle branch block
Posterior infarct (dominant R waves in V1 + ST↓ V1-3)

Contraindications

Absolute

Previous haemorrhagic stroke
Other CVA <6 months
Active internal bleeding (menses excluded)
Known/ suspected aortic dissection

Relative

Severe uncontrolled hypertension
Known bleeding disorder/ on warfarin INR >2.5
Recent major surgery/ trauma/ internal bleeding < 4/52
Active peptic ulcer
Pregnancy
Previous allergic reaction/ streptokinase >4/7
CPR

Drugs available

Streptokinase

Pros: Inexpensive

Cons: May cause allergy/ anaphylaxis,
Avoid if previous use

Alteplase/ Retaplase

Pros: More likely to achieve thrombolysis

Cons: Expensive

Percutaneous transluminal coronary angioplasty (PTCA)

This can be primary or secondary, following thrombolysis. A guidewire is passed through the stenosis identified on angiography and a balloon inflated to reopen the artery (angioplasty). A stent may then be put in place to reduce risk of restenosis.

Pros: Successful procedure with good, proven results
Provides evidence of stenosis and reperfusion

Cons: Requires specialist staff and equipment, invasive, expensive

Coronary Artery Bypass Graft

This cannot be achieved sufficiently rapidly in an emergency situation to rescue ischaemic myocardium and carries a high risk. It may be considered once the patient is stable.

2. Unstable angina and NSTEMIs

These are often initially indistinguishable and are treated similarly. Patients should be given 300mg of clopidogrel stat (x1 year 75mg od) on top of the usual 300mg of aspirin (lifelong 75mg od). A β-blocker (lifelong) and LMW heparin is also indicated (e.g. 1mg/kg clexane bd x 48 hours). If pain continues, GTN infusion should be started.

Patients risk should be quantified as information becomes available. Patients fall into low (0-2) and high (3+) TIMI* risk according to:

- Age >65 years
- 3 risk factors for coronary artery disease
- 2 anginal episodes in 24 hours before presentation
- Use of aspirin in 7 days before presentation
- ST segment change >0.5 mm
- Elevated serum concentration of cardiac markers
- 50% coronary stenosis on angiography

*TIMI= Thrombolysis In Myocardial Infarction study, 2000

Those at high risk, with dynamic ECG changes or with positive troponin benefit from angiography as an inpatient. A glycoprotein IIb/IIIa inhibitor is given to patients awaiting emergency angiogram. Those at low risk or negative troponin should be managed conservatively and a stress test should be performed to determine their need for angiography.

Secondary prevention

Treatment is as for ischaemic heart disease (discussed above).

Aspirin + statin + β-blocker + ACE ± anti-anginals

Prognosis

Characteristics associated with a poor prognosis include increasing age, male sex, previous MI, diabetes, hypertension, and multiple-vessel or left main stem disease. Patients who survive to reach hospital have a mortality of about 8%.

Complications of myocardial infarction

Ischemic	Recurrent infarction
	Post-infarction angina
Mechanical	Cardiogenic shock and death
	Heart failure
	Ventricular septal rupture
	Papillary muscle rupture
	Mitral regurgitation
	Pseudoaneurysm
	Ventricular aneurysm

Arrhythmic	Ventricular ectopics Ventricular fibrillation Bradycardias Bundle branch block
Embolic	Mural thrombus embolisation Stroke/ renal infarction Limb/ intestinal ischaemia
Inflammatory	Pericarditis = Dressler's syndrome More likely if transmural MI Occurs 24-96 hours post MI ?Autoimmune aetiology

VALVULAR HEART DISEASE

HINTS FOR UNDERSTANDING VALVE DISEASE:

1. Severity

a. The loudness of the murmur does *not* correlate with disease severity

b. Disease severity can be assessed clinically by presence of:

Symptoms/ associated signs/ heart failure

c. Ultimately, the best measure of severity is **investigation** with an echocardiogram to determine:

Valve **structure**

Valve **function**

Heart function, looking to see if there is any **heart failure** (ejection fraction <45%)

2. Symptoms and signs

The best way to remember the signs and symptoms is to understand the *effect* of the lesion, looking at both pressure and volume effects. Increased **volume** results in **dilatation** which is an increase in the *size* of a chamber, while increased **pressure** results in **hypertrophy** which is an increase in the *wall thickness* (muscle) of a chamber. Only dilatation of the left ventricle results in a displaced apex. In hypertrophy, the apex is in the same place but the surrounding muscle is thicker.

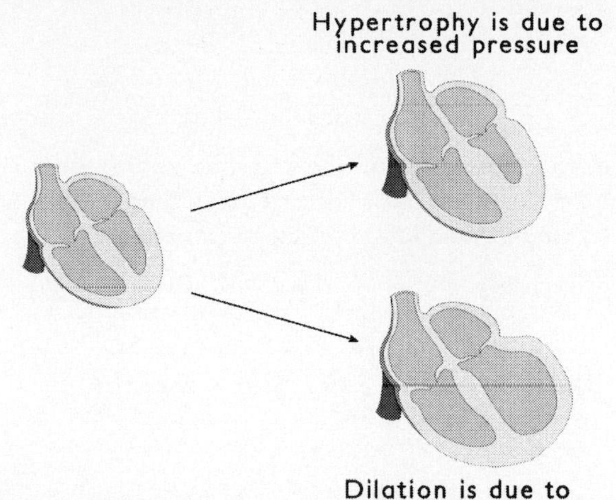
Hypertrophy is due to
increased pressure
Dilation is due to
increased volume

a. In *stenotic* valves, blood will be unable to be pumped forward, resulting in accumulation of blood *behind* that valve and an increase in **pressure** of the corresponding chamber and thus hypertrophy. Thus, either the ventricle in aortic stenosis or the atrium in mitral stenosis will hypertrophy. In mitral stenosis, the left atrium will dilate due to the increased **volume** of blood. In aortic stenosis, the left ventricle won't dilate, as the increased volume can initially be accommodated in this larger chamber.

b. In *regurgitant* valves, not all the blood flows in the right direction. A higher stroke volume has to be produced by the **left ventricle** because some of it will flow back during diastole and not contribute to the circulation. In mitral regurgitation, blood flows back into the left atrium, causing volume overload and dilatation. In aortic regurgitation, blood flows back into the left ventricle from the aorta. This increases **volume** in the left ventricle as a large stroke volume is needed to maintain a normal cardiac output and strains the heart towards failure. Hint: A displaced apex (due to ↑ LV volume) is therefore seen in *regurgitant* valve lesions and left heart failure.

c. An increase in **pressure** can also backtrack along the pump circuit, for example an increase in left ventricular pressure can raise left atrial pressure which increases pulmonary venous pressure, in turn raising pulmonary arterial pressure (*pulmonary hypertension*), which raises right ventricle pressure (*congestive heart failure* = right heart failure as a result of left heart failure). **Heart failure** is the endpoint of all valve lesions. Remember this is all just a **circuit** and you can work out all the effects.

What follows is a description of the four main valve diseases of the left heart, then a brief description of valve failure in the right heart.

Mitral Stenosis

Definition

Mitral stenosis is the narrowing of the opening of the mitral valve, almost always as a result of rheumatic heart disease, resulting in obstruction to blood flow from the left atrium to the left ventricle during diastole.

Description

Incidence	Rarer now, due to ↓ incidence of rheumatic heart disease
Age	↑ with age, seen in children if severe
Sex	F>M

Aetiology

Congenital	(rare)
Acquired	Rheumatic heart disease (mcc, 50%) Age related degeneration, fibrosis and calcification Prosthetic valve

Pathophysiology

Valve thickening, cusp fusion and calcium deposition lead to:

Narrowed opening
Immobility of cusps

Blood cannot pass into the left ventricle, increasing left atrium volume and pressure.

Mitral Stenosis

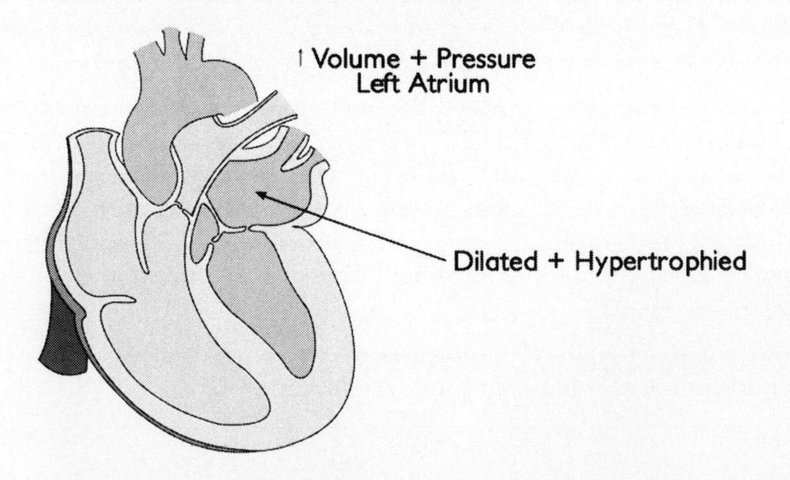

↑ Volume: Left atrial dilatation

→*atrial fibrillation*

↑ Pressure: Left atrium hypertrophy*:

↑ pulmonary venous pressure

↑ pulmonary arterial pressure

→*Pulmonary hypertension*

Pulmonary oedema

Pulmonary regurgitation

→***Right heart failure***

Mitral stenosis is the *only* one of the left sided valve disorders that does *not* directly cause left heart failure, but rather **right heart failure**. The other disorders result in left heart failure which may lead to congestive heart failure (RHF as a result of LHF). *Note: hypertrophy is theoretical and not really seen, (e.g. on echo) as atrial wall is so thin-walled to begin with.

History and Examination

CAUSE: PMH of rheumatic heart disease/ prosthetic valve

EFFECT:

Symptoms are not seen until moderate stenosis is present ($2cm^2$ instead of the normal $5cm^2$ valve orifice), and are the result of complications.

Atrial fibrillation	Palpitations
	Systemic emboli (mc: stroke)
Pulmonary oedema	Pink frothy sputum, bronchitis
Right heart failure	Leg oedema

Signs include:

Face	Mitral facies =dusky flushed cheeks (cyanosis)
Pulse	AF, small volume
JVP	Raised JVP =right heart failure
Apex	Tapping, undisplaced
Mitral area	Diastolic thrill
Heart sounds	Loud S1=closing of stenotic mitral valve
	Opening snap =sudden opening at ↑ pressure
	Loud P2 =pulmonary hypertension

Murmur	"<u>**Mid-diastolic rumble**</u>"
	Best heard: Over the apex In expiration With the patient in the left lateral position Radiating to the axilla
	±Graham-Steell murmur =pulmonary regurgitation secondary to pulmonary hypertension

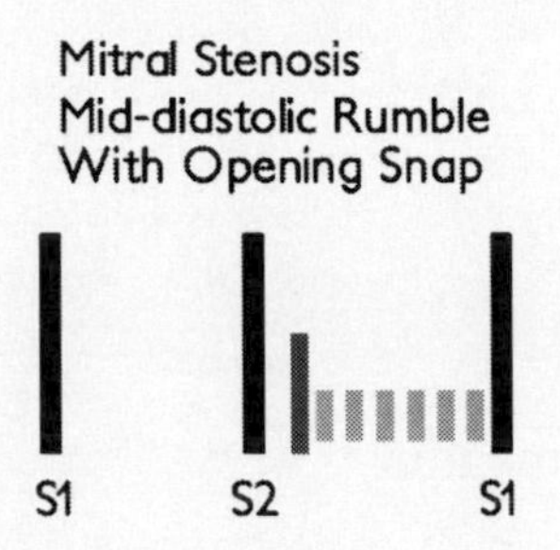

Severity can be assessed by:

Presence of associated signs

Mitral facies, AF, tapping apex

Quality of murmur (*not* grade)

Opening snap closer to S2
Longer diastolic murmur

Pulmonary hypertension

RV heave, loud P2, GS murmur

Right heart failure

↑JVP
Peripheral oedema
Hepatomegaly
Ascites

Most patients do not require treatment, however treatment options include:

Symptomatic	Dyspnoea	Diuretics, treat bronchitis
Definitive	Pulmonary HTN	Surgical intervention
		=valvotomy/ replacement

Atrial fibrillation should be treated and endocarditis prophylaxis given for surgical procedures.

Aortic Stenosis

Definition

Aortic stenosis is the narrowing of the opening of the aortic valve, resulting in obstruction to blood flow from the left ventricle to the aorta during systole.

Description

Incidence	Very common murmur, 10% prevalence in 80 year olds
Age	most commonly seen in the elderly
Sex	M>F in congenital AS

Aetiology

Congenital	Turbulent blood flow through bicuspid aortic valve as these tend to degenerate faster
Acquired	Age related degeneration, fibrosis and calcification (mcc)
	Rheumatic heart disease (mitral stenosis also likely to be present)
	Prosthetic valve

Pathophysiology

Valve thickening, cusp fusion and calcium deposition lead to:

Narrowed opening

Immobility of cusps

Blood cannot pass into the aorta, increasing left ventricle pressure.

↑ Pressure:

Left ventricle hypertrophy which can result in relative ischaemia:

→angina

→arrhythmias

→***Left heart failure***

Aortic Stenosis

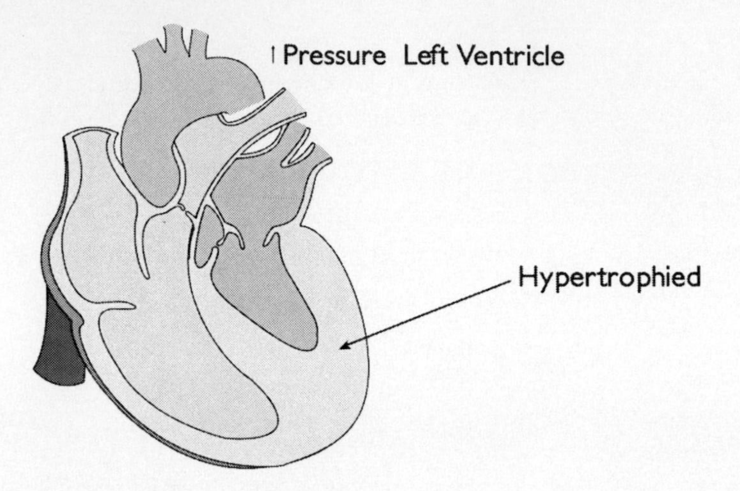

<u>History and Examination</u>

CAUSE: PMH of rheumatic heart disease/ prosthetic valve

EFFECT:

Symptoms are not seen until moderate stenosis is present (1/3 of normal valve orifice), and are the result of the heart's inability to increase cardiac output.

Exercise induced syncope

Angina

Left heart failure→ dyspnoea, orthopnoea, PND

Prognosis is poor if symptoms are present: death within 3 years if no intervention.

Signs include:

Pulse	Slow rising, small volume, sinus rhythm
BP	Narrow pulse pressure
Apex	Heaving, undisplaced
Heart sounds	
	S4= atrial contraction against stiff ventricle
	"Ten-nes-see"
	Soft A2 = immobile aortic valve
	Ejection click = sudden opening of bicuspid valve

Murmur “**<u>Ejection systolic murmur</u>**”

Best heard:
Aortic area
In expiration
With the patient leaning forward
Radiating to the neck

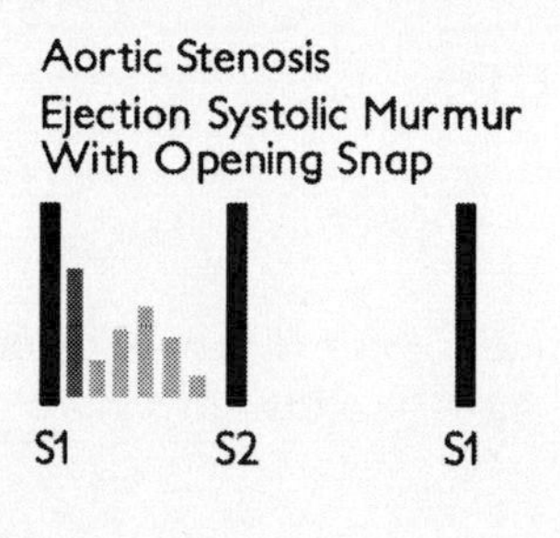

Severity can be assessed by:

Presence of associated signs
Slow rising pulse, narrow pulse pressure, heaving apex

Quality of murmur (*not* grade)
Longer systolic murmur
Loss of A2

Left heart failure
Bibasal crepitations lung bases

Congestive heart failure (RHF as a result of LHF)
↑JVP, peripheral oedema, hepatomegaly, ascites

Symptoms are a good indication of severity and therefore need for Rx:

Asymptomatic	Incidental finding	→ regular follow up
Symptomatic	Syncope, angina	→ valve replacement
Severe AS	>100mmHg pressure gradient	→ valve replacement

Endocarditis prophylaxis is essential.

AORTIC SCLEROSIS:

This is calcification and thickening of the aortic valve with no obstruction to ventricular outflow.

Incidence increases with age, male gender, smoking, hypertension, high LDL and diabetes. As a potential marker of coexisting coronary disease, it is associated with an increase of ~50% in risk of death from MI. Aggressive management of modifiable risk factors may slow its progression.

Compared to aortic stenosis, the carotid pulse is normal in volume, the ejection systolic murmur is softer and S2 is normal/loud, but it is best differentiated by echocardiography.

MITRAL REGURGITATION

Definition

Mitral regurgitation is the failure of mitral valve closure resulting in reflux of blood from the left ventricle back into the left atrium during systole.

Description

Incidence Very common

Age May present at any age due to its various causes
Though ↑ with age

Aetiology

The many causes of mitral regurgitation fall into two categories:

Intrinsic valve disease

Rheumatic heart disease (mcc -50%)
Prolapsing mitral valve*
Endocarditis
Ruptured chordae tendinae: MI, trauma, infective endocarditis

Functional (stretched ring)

Dilated LV — Ischaemic/ hypertensive heart disease
Aortic regurgitation/ stenosis
Dilated cardiomyopathy
Collagen disorders: Marfan's/ Ehlers-Danlos

Pathophysiology

The mitral valve is unable to close completely due to disease or stretching which leads to reflux of blood from the left ventricle back into the left atrium during systole.

↑ Volume: Left atrium dilatation to accommodate ↑ blood volume

(Effects of this are seen late in disease)

→***Atrial fibrillation***
→***Pulmonary hypertension***
→***Congestive heart failure***

↑ Pressure: Left ventricle hypertrophy and later dilatation due to ↑ volume to ↑ stroke volume to maintain CO

→***Left heart failure***

Mitral Regurgitation

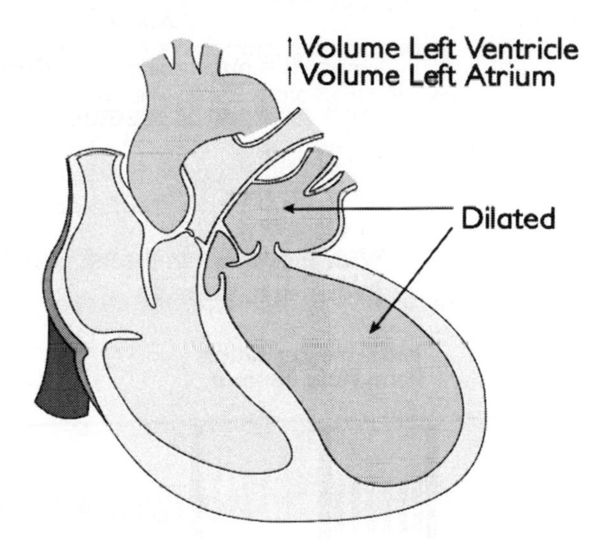

History and Examination

CAUSE: PMH of rheumatic heart disease, endocarditis

EFFECT:

Symptoms are not seen until late in disease when the left ventricle can no longer compensate for the increased requirement in stroke volume, or when pulmonary hypertension occurs.

↑ Stroke volume	Palpitations
↓ Cardiac output	Fatigue, lethargy
Atrial fibrillation	Palpitations

Intrinsic valve disease

Rheumatic heart disease (mc)
Endocarditis (mc)
Bicuspid aortic valve
Ruptured chordae tendinae: MI, trauma, infective endocarditis

Functional (stretched ring)

Dilated aortic root: Hypertension
Aortic dissection
Collagen disorders: Marfan's/ Ehlers-Danlos
Seronegative arthritides
Syphilis

Pathophysiology

The aortic valve is unable to close completely due to disease or stretching which leads to reflux of blood from the aorta back into the left ventricle during diastole.

↑ Volume:

Left ventricle dilatation to accommodate ↑ blood volume
→***Left heart failure***
→***Congestive heart failure***

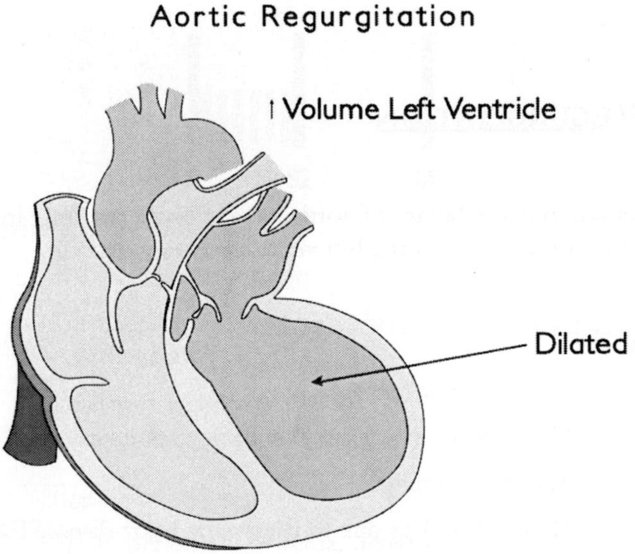

Aortic Regurgitation

↑ Volume Left Ventricle

Dilated

History and Examination

CAUSE: PMH of underlying disease, mc: rheumatic heart disease, endocarditis

EFFECT:

Symptoms are not seen until late in disease when the left ventricle can no longer compensate for the increased requirement in stroke volume.

↑ Stroke volume	Palpitations
↓ Cardiac output	Fatigue, lethargy
Left heart failure	Dyspnoea, orthopnoea, PND

Signs include:

Pulse	Sinus rhythm, collapsing pulse	
BP	Wide pulse pressure	
Apex	Thrusting, displaced, systolic thrill	
Heart sounds	Soft S2	= incomplete closing of aortic valve
	S3	= volume overloaded ventricle
		"*Ken-tuc-ky*"
Murmur	"**<u>Early</u> <u>diastolic</u>**"	
	Best heard:	

Left sternal edge + aortic region
In expiration
With the patient leaning forward
Radiating to the neck

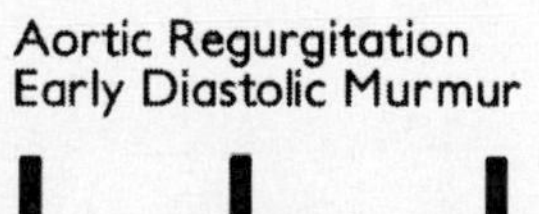

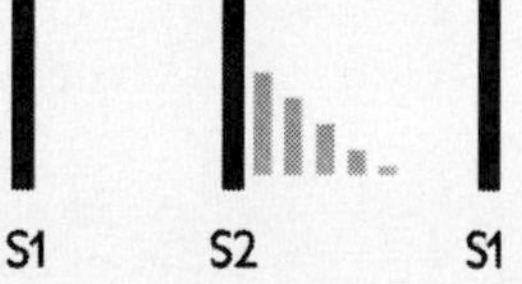

±Austin Flint murmur
=regurgitant jet hitting anterior MV cusp
Mid-diastolic murmur (similar to MS)

Signs of hyperdynamic circulation (uncommon)

Corrigan's	Visible carotid pulsations
Quinke's	Capillary pulsation in nail beds
De Musset's	Head nodding with systole
Duroziez's	Femoral diastolic murmur due to backflow
Traube's	Pistol shot sound over femoral arteries

Severity can be assessed by:

Presence of associated signs

Collapsing pulse, wide pulse pressure
Displaced apex: the more severe, the larger the LV
Signs of hyperdynamic circulation

Left heart failure

Bibasal crepitations

Evidence of LV enlargement is a good indication of severity, symptoms occur late. Therefore *Symptoms/ LV enlargement* = valve replacement.

Most patients will require valve replacement.
Endocarditis prophylaxis should be given.

Investigation and Treatment of valve disorders

Investigations

The aim of investigation is to:

Diagnose valve lesion
Assess severity
Determine cause and complications

e.g. Rheumatic fever (cause)
Endocarditis (cause and complication)
Heart failure (complication)

Simple	ECG	**LVH** in all except mitral stenosis **Bifid p waves** in MS and MR =atrial delay due to enlarged atrium "P-mitrale" **Arrhythmias**, especially AF in MS/MR
	Urine dipstick	Endocarditis (haematuria)

Bloods FBC, CRP, ESR — Endocarditis

Imaging **CXR**

Size of heart chambers

MS	Small heart, ↑ LA
AS	Small heart (initially) Dilated aorta
MR	↑ LV
AR	↑ LV

Calcified valves (white on CXR)

Evidence of heart failure (*ABCDEF*)

Perihilar **A**lveolar oedema
Kerley **B** lines
Cardiomegaly
Upper lobe blood **D**iversion
Pleural **E**ffusion
Fluid in **F**issures

Echocardiogram is the gold standard for valvular lesions:

1. Valve **structure**

   Congenital abnormalities
   Endocarditis

2. Valve **function =** degree of stenosis/ reflux

   Stenosis: pressure gradients across valve
   Regurgitation: regurgitant jet measured

3. Chamber **size**

   Mitral Stenosis =obstruction to flow, accumulation **LA**
   High pressure and volume in LA: **LA hypertrophy + dilatation**
   →RVF

   Aortic Stenosis =obstruction to flow, accumulation **LV**
   High pressure in LV: **LV hypertrophy**
   →LVF →RVF*
   Note no change in size as LV can accommodate initial ↑ volume

   Mitral Regurgitation =regurgitation to **LA**, more work for **LV**
   High volume in LA: **LA dilatation**
   High volume in LV: **LV dilatation**
   →LVF →RVF*

   Aortic Regurgitation =regurgitation to **LV**, more work for **LV**
   High pressure and volume in LV: **LV hypertrophy + dilatation**
   →LVF →RVF*

   *Remember that LVF can lead to RVF (congestive heart failure)

4. Chamber **function** = ejection fraction

Measures **LV systolic function** and quantifies degree of heart failure
except in MS: RV systolic function

Note that most echoes are done with the probe on the anterior chest wall (*transthoracic echo*). An alternative is *Transoesophageal echo* (TOE), which provides better visualisation of posterior heart chambers (LA, LV), endocarditic vegetations and mechanical valves.

Cardiac catheterisation

This is most commonly used for angiography, but a small BP monitor attached to the end of the probe allows measurements either side of the valves. Alternatively, by injecting dye into the heart chambers themselves (rather than the coronary arteries as in angiography), flow and ventricular function can be measured.

Stenosis: pressure gradients across valve
Regurgitation: regurgitant jet visualised

Cardiac catheterisation is not often necessary as most of the information can be obtained from the echocardiogram which is non-invasive. It is most useful in aortic stenosis though generally usually done to look for ischaemic heart disease rather than to assess the valve gradient.

Treatment

CAUSE: Treat underlying cause (especially in AR/MR)
Treat cardiac risk factors (especially in AS)

EFFECT:

Conservative
Follow up with echocardiograms

Medical
Endocarditis prophylaxis
Treatment of AF
Treatment of heart failure/ pulmonary hypertension

Interventional

Indications

Not all stenosed valves need surgical intervention. Indications include:

MS Pulmonary hypertension, persistent dyspnoea
AS Symptoms (syncope, angina)

Treatment options include:

Valvuloplasty	Useful in MS or those unsuitable for surgery
Open valvotomy	Cardiopulmonary bypass required
Valve repair	Mostly useful in MR
Valve replacement	For calcified/ regurgitant valves

Most symptomatic AS requires valve replacement. Most regurgitant valves need surgical intervention.

MR	LV enlargement, symptoms	→repair/replace
AR	LV enlargement, symptoms	→replace

Valve replacement is the definitive treatment for any valve disease. Options include:

Tissue

Porcine/bovine = xenograft
Cadaveric = allograft
Patient's pulmonary valve = autograft

Last ~10 years, don't require anticoagulation
Good in older patients with a shorter lifespan and a higher falls + bleed risk.

Mechanical

Ball and cage	*1 metallic click*
Tilting disc	*2 metallic clicks*
Double tilting disc	

Longer lasting, require anticoagulation

Complications (long term)

Infection → endocarditis
Detachment → valve failure
Thrombosis and embolism → stroke
Haemorrhage secondary to anticoagulation
Haemolysis → anaemia

Prognosis

Prognosis varies across the valve diseases and individual patients. It correlates with the severity of disease and complications such as heart failure.

RIGHT HEART VALVE DISEASE

Due to its rarity, right sided valve disease is nowhere near as important for exams as left sided disease. The pathophysiology is similar to left sided disease, with right heart failure as the common end-point. Murmurs are best heard during inspiration.

Tricuspid stenosis: Usually due to rheumatic heart disease F>M
Rarely occurs in isolation.

Pulmonary stenosis: Usually congenital: Tetralogy of Fallot /Rubella

Tricuspid regurgitation:

Functional: Dilated RV/ pulmonary HTN
Organic: Rheumatic heart disease
Endocarditis, esp: IVDU, staph aureus

Pulmonary regurgitation:
Pulmonary hypertension (Graham-Steell murmur)

RHEUMATIC FEVER

Definition

An increasingly rare inflammatory disease in children; it is a complication of group A streptococcal infection. It affects the heart, skin, joints and CNS.

Description

Age Children usually between 5-15 years of age

Geography Common in Middle East and South America

Aetiology

Rheumatic Fever is one of the sequelae of group A streptococcal infection, occurring 2-3 weeks after, usually an URTI.

Pathophysiology

Micropathology

Antibodies against streptococcus react to normal valve tissue and other affected tissues in the body. This is thought to be due to "molecular mimicry" and results in the "Aschoff nodule", a granuloma.

Macropathology

Warty vegetations grow on valves, pericarditis with serofibrinous effusion

History and Examination

CAUSE: PMH streptococcal infection (sore throat)

EFFECT: Sudden onset fever, arthritis, malaise

Jones' Criteria = 2 major OR 1 major + 1 minor criteria
+ Evidence of strep infection

Major

1. Carditis	New murmur, pericardial effusion, pericarditis
2. Polyarthritis	Large joints, no permanent damage
3. Sydenham's chorea	Semi-purposeful movements
4. Erythema marginatum	Macular rash with raised edges
5. Subcutaneous nodules	Small, painless, joint extensor surfaces

Minor

1. Fever
2. Arthralgia
3. ↑ CRP/ESR
4. ↑WBC
5. ↑ PR interval on ECG

Evidence of streptococcal infection

Throat cultures
Antistreptolysin O (ASO) antibody titres
Recent scarlet fever

Investigations

Simple	**ECG**	Pericarditis	Saddle-shaped ST segment
		Myocarditis	Inverted/flattened T waves
		Arrhythmias	
	Throat swab		
Blood	**FBC**		
	ESR/CRP		
	ASO Ab titre		

Treatment

Conservative

Bed rest for acute phase

Medical

Penicillin
High dose aspirin for arthritis (limited by tinnitus= toxicity)

Prophylaxis:	Penicillin until no longer at risk of strep infection (~30 yrs old)
	Thereafter give with dental work / surgery

Prognosis

50% with carditis develop chronic rheumatic valve disease

Mitral	50%
Mitral + aortic	40%
Aortic alone	2%

Initially regurgitation occurs, which is followed by valve stenosis years later.

INFECTIVE ENDOCARDITIS

Definition

Infective endocarditis is an infection of the endocardium, (the inner lining of the heart). Previously known as "subacute bacterial endocarditis", the term 'infective' is now preferred in recognition of the fact that it can be either acute or subacute, and that it can have a fungal aetiology. 'Infective' also differentiates it from inflammatory endocarditis.

Description

Incidence	Relatively uncommon in developed countries
Age	↑ with age, as higher prevalence of damaged valves
Geography	Increased incidence in developing countries

Aetiology

Infective endocarditis occurs as a result of:

1. **Bacteraemia**

    Dental work
    Surgery
    UTI
    URTI
    IVDU =normal tricuspid valve is usually affected
    Non-infective endocarditis can be seen in SLE and malignancy.

2. **Abnormal valve/ endocardium**

Mitral/ aortic valve disease	Especially regurgitant lesions
Rheumatic heart disease	
Congenital heart disease	PDA/ VSD/ coarctation
Prosthetic valves	

Pathophysiology

Micro

A multitude of organisms have been implicated in endocarditis. These are the more common:

Streptococcus viridans (mcc)

Pharynx: URTI
Dental work

Enterococcus faecalis

Perineum, faecal UTI

Staphylococcus aureus

Skin: Catheters/ CVP line
Cellulitis/ abscess
This is usually a very aggressive infection

Staphylococcus epidermidis

Skin: IVDU
This is usually a more indolent infection

HACEK is an all-inclusive term for endocarditis due to *Haemophilus, Actinobacillus, Cardiobacterium, Eikenella,* and *Kingella* species of bacteria. Clinically, these are characterised by a subacute or chronic course, and often present with embolic lesions from large vegetations.

Fungal: Candida, aspergillus, histoplasma

IVDU and patients with prosthetic valves are more susceptible to Staphylococcus epidermidis and fungal infection.

Macro

1. Destruction of the valve leaflet can result in a regurgitant lesion.

2. Vegetation formation = platelets, fibrin and infecting organism. These may embolise, particularly if acute with large vegetations.

3. Extracardiac manifestations are due to:
   Embolisation
   Immune complex deposition e.g. in kidneys

History and Examination

Fever + new murmur = endocarditis until proven otherwise.

CAUSE:

Risk factors	Bacteraemia, damaged valve

EFFECT:

Infection	Fever, malaise, myalgia
Cardiac disease	New murmur, heart failure

Immune complexes and microemboli

Roth spots	Small retinal haemorrhages
Janeway lesions	Macules on thenar eminence
Osler's nodes	Hard, tender, on fingers, toes
Splinter haemorrhages	Subungal haemorrhages

Haematuria from glomerulonephritis/ renal emboli

Petechial haemorrhages

Clubbing

Emboli Stroke, MI, PE, Splenic infarct

Investigations

Simple	Urine dipstick	Haematuria
	MSU	Microscopic haematuria, organism
	ECG	MI, conduction defects
Blood	**Blood cultures**	3 sets within 24 hours Hopefully before ABx given
	FBC	Anaemia-normochromic/cytic Leukocytosis Thrombocytopenia
	CRP, ESR	Both are raised
Imaging	**Echocardiogram**	Vegetations Regurgitant valve lesion
	CXR	Heart failure, PE

The diagnosis can be made on Duke's criteria (don't confuse these with Jones' criteria for Rheumatic Fever!)

2 major
1 major + 3 minor
5 minor

Major

1. **2 x positive blood cultures**
2. **Evidence of endocardium involvement**

   Positive echocardiogram: vegetations/abscess
   New regurgitation

Minor

1. Risk factors: bacteraemia/ damaged valve
2. Fever
3. Immune complex/ microembolic signs
4. Positive blood cultures that don't meet the major criteria
5. Positive echocardiogram not meeting the major criteria

Treatment

Conservative

This is a serious disease that if left untreated has an almost 100% mortality! Aggressive medical, and if necessary, surgical treatment should follow.

Medical

Antibiotics

Consult with microbiology early

Empirical treatment: i.v. benzylpenicillin and gentamicin
Fluconazole for fungal infections

Change to definitive treatment on growth of an organism

Interventional

Surgical → valve replacement

Indications: Valve rupture
Heart failure
Resistant infection, especially if prosthetic valve

Preventative

At risk patients get prophylaxis for certain procedures:

Patients: Prosthetic valve, previous endocarditis
Valve disease, septal defects

Procedures: Dental work
→ Amoxicillin

Lower GI/GU instrumentation
→ Amoxicillin + gentamicin

Prognosis

Mortality can still be over 30% with antibiotic and surgical treatment, though it depends on the organism. Staphylococci are the most virulent. The prognosis is worse if the patient is in heart failure, has a prosthetic valve or the organism is not isolated.

MYOCARDIAL DISEASE

These are all uncommon diseases and are less likely to feature in exams.

Myocarditis =Inflammation of cardiac muscle

Cardiomyopathy =Idiopathic diseases of cardiac muscle
- Dilated
- Hypertrophic
- Restrictive

MYOCARDITIS

Definition: Inflammation of the myocardium

Aetiology:

Infection

Virus	Coxsackie (mcc), influenza, rubella, HIV
Bacteria	Diphtheria toxin
Protozoa	Trypanosoma cruzi

Inflammation

Radiation	
Drugs	Hypersensitivity reaction to: Lead poisoning, chloroquine etc
Autoimmune	Rheumatic fever

History and Examination: Acute illness + cardiac failure

Management: Diagnosed on clinical suspicion, though definitively on biopsy. Treat underlying cause.

CARDIOMYOPATHIES

Dilated Cardiomyopathy

Definition: *Idiopathic* dilatation of the ventricles that may result in heart failure.

Risk factors: Family history (20%), alcohol, thyrotoxicosis, autoimmune disease.

History and Examination: Congestive heart failure, arrhythmias.

Management: Diagnosed on echocardiogram. Treat heart failure, arrhythmias, consider heart transplant.

Hypertrophic Cardiomyopathy / Hypertrophic Obstructive Cardiomyopathy

Definition: *Idiopathic*, asymmetrical hypertrophy of the myocardium, typically of the septum, that may result in LV outflow obstruction (HOCM) and heart failure.

Aetiology: The majority of cases are autosomal dominant: FH of sudden death.

History and Examination: Angina, syncope, arrhythmias, heart failure.

Management: Diagnosed on echo. Treat with β-blockers; implantable defibrillator if at risk of sudden death.

Restrictive Cardiomyopathy

Definition: *Idiopathic* stiffening of the myocardium, resulting in poor ventricular filling and heart failure.

Aetiology: Infiltrative disease: amyloidosis (mcc), sarcoidosis, haemochromatosis.

History and Examination: Heart failure, embolisms + signs similar to constrictive pericarditis.

Management: Diagnosed on echo. No specific treatment, consider heart transplant.

PERICARDIAL DISEASE

Acute pericarditis
Pericardial effusion / tamponade
Constrictive pericarditis

These conditions are interrelated, unlike myocardial diseases. Acute pericarditis can lead to pericardial effusion. If the effusion is large enough, it may cause cardiac tamponade. Constrictive pericarditis is also one of the sequelae of acute pericarditis.

ACUTE PERICARDITIS

Definition Inflammation of the pericardium

Aetiology

Viral	(mc in younger pts) Coxsackie; seen in young adults Good prognosis, may recur/cause sudden death
Post-MI	(mc in older pts) Dressler's syndrome occurs 2-10 weeks after MI
Autoimmune	Rheumatic fever, RA, SLE, hypothyroidism
Malignant	Lung CA, breast CA, Hodgkin's lymphoma
TB	Frequently results in constrictive pericarditis
Bacterial	Rare and fatal: Staph, H. influenzae
Uraemia	Terminal stages of renal failure
Drugs	Hydralazine, isoniazid, cyclophosphomide
Radiation	

History and Examination

Symptoms:

Chest pain

Substernal, sharp
Worse on inspiration, movement, coughing, lying flat
Relieved sitting forward

There may be a fever if due to infection or infarction.

Signs:

Pericardial friction rub

Superficial, leathery sound
Best heard at lower left sternal edge with patient sitting forward

Investigations

The aim of investigation is to diagnose pericarditis, exclude ACS and establish cause. ECG is diagnostic.

Simple	**ECG**	Saddle-shaped ST elevation T wave inversion (later)
Blood	FBC	Leukocytosis
	U&E	Uraemia
	Cardiac enzymes	MI
	Autoantibodies	RF, RA, SLE
	TFT	Hypothyroidism
	Serology	Viral
	Blood cultures	
Imaging	CXR	-cardiomegaly if pericardial effusion
	Echo	-if suspected pericardial effusion

Treatment

The aim of treatment is to treat the cause and alleviate symptoms

Conservative

Bed rest

Medical

Treat cause
Treat pain: ibuprofen
High dose aspirin
Systemic steroids if severe

PERICARDIAL EFFUSION

Definition A collection of fluid in the pericardial sac which may lead to cardiac tamponade as intracardiac pressure rises. **Cardiac tamponade is a *medical emergency*** as ventricular filling is poor and therefore cardiac output is compromised.

Aetiology Causes are as for pericarditis.

History and Examination

Signs and symptoms are as for pericarditis. As the effusion collects, the apex beat is harder to feel and heart sounds become soft and muffled, eventually, the pericardial rub is no longer heard.

Look out for cardiac tamponade (see below)

Investigations

The aim of investigation is to diagnose the pericardial effusion and exclude cardiac tamponade.

Simple	ECG	Low voltage QRS
Imaging	CXR **Echocardiogram**	Enlarged, globular heart Echo-free zone around heart
Invasive	Pericardiocentesis	mc&s, cytology ZN stain/ TB culture

Treatment

The aim of treatment is to treat the pericarditis as above and prevent cardiac tamponade.

CARDIAC TAMPONADE

Definition

Cardiac tamponade is a *medical emergency* that occurs as a result of fluid accumulating in the pericardial sac, restricting ventricular filling and compromising cardiac output.

Aetiology

Pericarditis→ pericardial effusion→ cardiac tamponade

Haemorrhagic i.e. blood rather than serous fluid fills the sac:
Trauma, usually blunt, e.g. RTA
Iatrogenic e.g. cardiac biopsy/catheterization
Aortic dissection
Warfarinisation

History and Examination

Beck's triad

↓ BP
↑ JVP (+ distended neck veins)
Muffled heart sounds

Kussmaul's sign: ↑ JVP on inspiration

Cardiac output may be so reduced that the patient goes into cardiogenic shock. Cardiac arrest may follow surprisingly quickly. Cardiac Tamponade is one of the four Ts of reversible causes of cardiac arrest!

Investigations

ECG changes are as for pericardial effusion (low voltage QRS)
An echocardiogram is diagnostic. There may not be time for one though.

Treatment

The ultimate treatment is to relieve intracardiac pressure by removing the pericardial fluid through **needle pericardiocentesis**. This is still an emergency, so start with ABC, especially if there is trauma as there can be multiple injuries.

A Airway + adjunct
B Breathing + oxygen
C Circulation + i.v. access + send bloods

Needle pericardiocentesis:

Wide bore needle + three way tap + 20ml syringe
Introduce at 45° below and to the left of xiphisternum
Aim for tip of left scapula
Aspirate continuously, watch ECG for changes
ST depression/ ectopic beats? = you're in the myocardium, withdraw

You know you're in the right place if:
The patient improves
The aspirate doesn't clot as it contains pericardial fluid
(if it clots, it was ventricular blood)

The aspirate should be sent off for mc&s, TB culture + ZN stain

Complications:

Aspiration of ventricular blood
Arrhythmias
Damage to anatomy:

Lacerated	Ventricle	
	Coronary artery	
Punctured	Lung	→pneumothorax
	Aorta	
	Oesophagus	→mediastinitis
	Peritoneum	→peritonitis

CONSTRICTIVE PERICARDITIS

Definition

The thickening, fibrosis and calcification of the pericardium following pericarditis. This results in limited ventricular filling as the heart is encased and thus constrained by a rigid pericardium.

Aetiology

Certain types of pericarditis are more likely to become constrictive:

TB
Bacterial
Rheumatic heart disease

History and Examination

Features of venous congestion, similar to right heart failure:

↑ JVP

Bilateral pitting leg oedema

Hepatomegaly

Kussmaul's sign: ↑ JVP on inspiration

Atrial fibrillation (enlarged RA)

"pericardial knock"

= loud diastolic 3rd heart sound of rapid ventricular filling

Investigations

ECG Low voltage QRS

CXR Small heart + *calcification*

Echo Thick pericardium

Cardiac catheterisation shows diastolic pressures equal in all 4 heart chambers.

Treatment

Treat cause, AF and heart failure. Surgical resection of myocardium helps 50%.

RESPIRATORY

RESPIRATORY

HISTORY

CAUSE

Respiratory risk factors

1. Genetics	Family history
2. Exposure	Smoking
Lung cancer	Smoking/ asbestos
Asthma	Cold/ exercise
	Allergens: pets/ dust/ occupation
TB	Contacts/ travel
PE/DVT	Immobility/ active cancer/ major surgery

EFFECT

Symptoms and ***severity***

1. Shortness of breath

Determine rapidity of onset and rate of progression

Sudden		PE, pneumothorax
Rapid	(mins-hours)	Asthma, pulmonary oedema
	(days-weeks)	Above + pneumonia, pleural effusion
Gradual	(months-years)	COPD, fibrosis, neuromuscular

Determine response to inhalers

SOB at night?

Orthopnoea?

Determine exercise tolerance

Medical Research Council Dyspnoea scale

1. SOB on strenuous exercise
2. SOB walking up a slight hill
3. Stops for breath when walking at own pace
4. Stops for breath after 100m / few minutes on level ground
5. Breathless when dressing/ undressing

2. Cough + sputum

Quantify:

Amount	Cupful/teaspoon
Colour	White/yellow/green
Haemoptysis	Cupful/teaspoon/tinge

3. Wheeze

Polyphonic: This is a result of collapse of many small airways on expiration – seen in asthma and COPD due to obstruction/ air trapping. The presence of wheeze indicates that the patient's peak expiratory flow rate is likely to be less than 50% of normal.

Monophonic: This is a result of stiffening of one large airway e.g. due to a tumour.

4. Chest pain

Pleuritic pain = sharp pain worse on inspiration; also elicit all the features of pain as described in the cardiology chapter.

5. Constitutional symptoms

Weight loss
Loss of appetite
Fatigue

EXAMINATION

Introduce

Name/age/occupation

Observe

1. **Surroundings** (Oxygen, sputum pot, inhalers)
2. **BMI** (cachectic?)
3. **Shortness of breath** (use of accessory muscles, lip pursing, sitting up)

Hands Clubbing, tar stains, peripheral cyanosis

Wrist Respiratory rate (do this as you pretend to take the pulse)

Pulse (bounding: CO_2 retention)

Hypertrophic pulmonary osteoarthropathy (HPOA)
= tender, swollen wrists (lung CA)

Arms CO_2 retention flap

Face **1. Eyes:** Anaemia (you only need to look in one eye)

Ptosis: Horner's due to apical lung tumour

2. Mouth: Central cyanosis (visible in mucosal membranes)

Neck

1. Trachea Central? If not, is it being pulled or pushed?

Tracheal tug = moves down on inspiration

2. Cervical lymphadenopathy

Palpate the neck from behind
(this may be done later, when the patient is sitting up)

3. JVP Cor pulmonale (RHF due to pulmonary hypertension)

Chest

1. **Observe** chest, axillae and back for scars (e.g. lobectomy)

Shape: pectus excavatum, barrel chested, kyphoscoliosis

2. **Palpate**

Chest expansion

Vocal fremitus

3. **Percuss** over chest, apices, and axillae

Resonant/dull "stony dull" implies pleural effusion

4. **Auscultate** over chest, apices, and axillae

1. **Intensity**

Reduced breath sounds? Does R=L?

2. **Character**

Vesicular breathing is "normal" breathing

Bronchial breathing is harsh: consolidation/fibrosis

3. **Added sounds** - listen carefully to character and timing

Wheeze (usually expiratory; due to airway obstruction)

Polyphonic = various airways (asthma/COPD)

Monophonic = one large airway (lung CA)

Stridor (inspiratory; due to obstructed large airway)

Crackles (inspiratory; due to re-opening of airways)

Fine = distal: atelectasis, pulmonary oedema

Coarse = proximal: bronchiectasis

Pleural rub

=Visceral and parietal pleura rubbing against each other

e.g. inflamed pleura due to lung infection/infarction

Back

Repeat **observation, palpation, percussion** and **auscultation** as for the chest. Some people choose to start with the back as most pathology is to be found here. It may also be easier to palpate for cervical lymphadenopathy at this point as the patient is sitting up and you can palpate the neck from behind which is the correct position. Look for sacral oedema.

Legs Pitting oedema, if present, look for level (ankle/knees/thighs)

"“To complete my examination, I would like…”

Temperature/ PEFR/ oxygen saturation

Putting signs together:

	Mediastinal shift*	**Percussion note**	**Breath sounds**	**Vocal resonance**	**Added sounds**
Consolidation	Pulled	Dull	Bronchial	↑	Crackles
Pleural Effusion	Pushed	Stony dull	↓	↓	Pleural rub
Pneumothorax	Pushed (if tension)	Hyper-resonant	↓	↓	None
COPD/ Asthma	Central	Hyper-resonant	Prolonged expiration	Normal	Wheeze

*See Trachea under Chest X-rays below

For the Thinking Medicine Respiratory Virtual Patient go to:

http://www.thinkingmedicine.com/elearning/resp/

Presentation

1. Name/ age/ occupation

2. Appearance Surroundings/ BMI/ SOB

e.g. "The patient was short of breath at rest, using accessory muscles and purse-lip breathing. He has a salmeterol inhaler."

3. Peripheral signs: Hands, wrist, arms, face, neck and legs

Give positives & relevant negatives

e.g. "He was not clubbed but his fingers were tar-stained. He has an increased respiratory rate of 18. He had no CO_2 retention flap, signs of anaemia or cyanosis. There was no cervical lymphadenopathy. His jugular venous pressure was not raised and there was no leg oedema."

4. Chest/back: Observation, palpation, percussion, auscultation

e.g. "On examination of his chest, there were no scars, he had poor expansion bilaterally and was hyperresonant to percussion. On auscultation, breath sounds were vesicular with a polyphonic expiratory wheeze."

5. Summary

Put everything together with positives & relevant negatives

e.g. "In conclusion, Mr Smith is a 79 year old smoker who has signs of chronic obstructive pulmonary disease as evidenced by his shortness of breath at rest, hyperinflated lungs with poor expansion and polyphonic wheeze, for which he uses a salmeterol inhaler. He has no signs of superimposed infection or cor pulmonale.

INVESTIGATION

CHEST X-RAYS

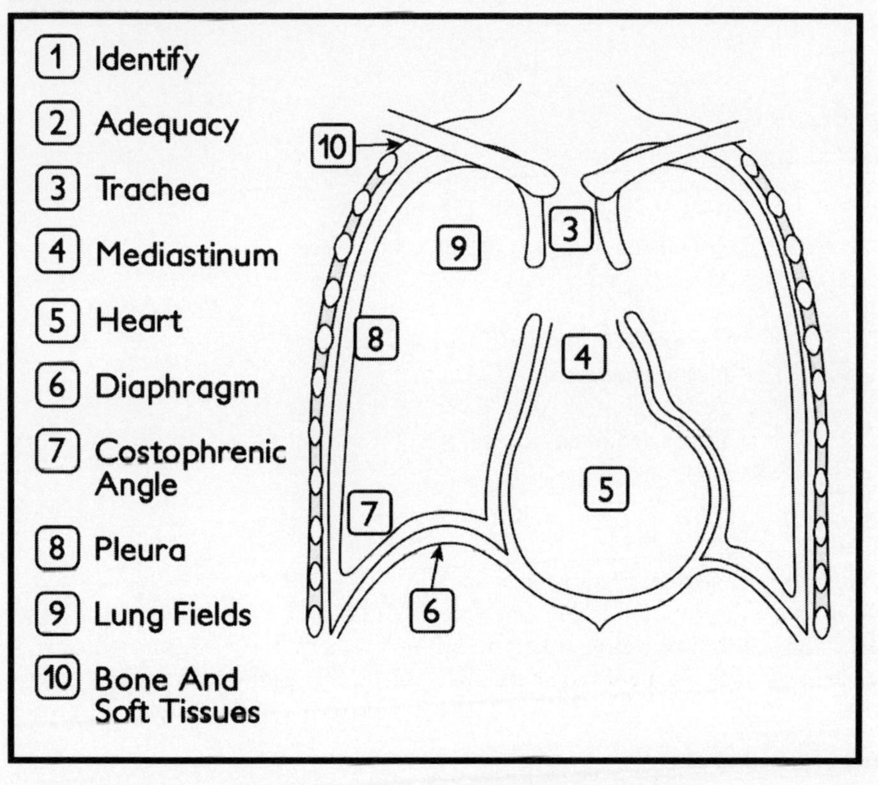

1. Describe

"This is an adequate AP/PA chest radiograph of whoever taken whenever" (Patient name, date taken, type of projection)

From here you can either go on to the most obvious abnormality, or work your way through all the points. In general, if there is a barn-door abnormality then head straight for it, but if it is something more subtle or indeed if you are not sure of any abnormalities, work through things systematically. Remember, you could be shown a completely normal x-ray, simply to test that you know how to interpret a film.

2. Adequacy

Only go into details if asked, otherwise it takes too much time

No rotation: spinous processes equidistant from clavicular heads
Good penetration: vertebrae faintly visible behind the heart
Adequate inspiration: at least 6 anterior ribs visible

3. Trachea

Central? If not, work out if it's being:

Pulled	Lobe collapse
Pushed	Tension pneumothorax Pleural effusion

4. Mediastinum

Shifted?

Pulled/pushed as above
2/3 of the heart should be on the left

Widened?

Lymphadenopathy: TB/sarcoidosis/lymphoma
Aortic aneurysm
Tumour
Thoracic goitre

5. Heart

Enlarged?

Normal cardio-thoracic ratio <50%
Note unable to comment if AP film
2/3 of heart should be on the on left
CCF/ pericardial effusion/ cardiomyopathy

Heart borders clear?

Right vertical	= Right atrium
Diaphragm	= Right ventricle
Left diagonal	= Left ventricle

6. Diaphragm

Flattened?

Hyperexpansion: COPD/asthma

Raised?

Lung volume loss, phrenic nerve palsy

Sub-diaphragmatic air?

Perforated viscus (ulcer/diverticulum) mcc
Post-surgical
Gynaecological, including water skiing

7. Costophrenic angle

Should be crisp and clear
If obliterated consider pleural effusion
-look for meniscus to confirm

8. Pleura Thickening/calcification/ masses/ pneumothorax

9. Lung fields

Zones:

Upper	Between apex and anterior insertion of 2nd rib
Mid	Between 2nd rib and anterior insertion of 4th rib
Lower	Anything below the anterior insertion of the 4th rib

Shadowing:

Radiographs show 5 densities:

Gas	Black
Fat	Dark grey
Fluid/soft tissue	Light grey
Bone/ calcification	White
Metal	Intense white

A line will only be seen at the interface between two densities.

Compare right and left lung fields, does one area look whiter/darker?

Too black = ↑ air / ↓ soft tissue

Hyperinflated lungs

> 8 *anterior* ribs visible
(Posterior ribs don't move as much on respiration)

No lung markings

If lung markings don't go all the way to the border:
Pneumothorax (look for edge of collapsed lung)

If wedge missing:

Pulmonary infarct (mcc PE)

Mastectomy

Too white (shadowing) = ↓ air or ↑ fluid / soft tissue

For shadows, describe:

Site Upper/middle/lower **zones** (not lobes)

Size

Character

a) Nodular shadows = circular = mass

Neoplasia	Single nodule: primary CA Multiple nodules: metastases
G*ranulomas*	Multiple, small: milliary TB Sarcoidosis, Wegener's

b) Ring shadows

Airways seen end-on (small)
Cavitating lesions e.g. abscess (TB/fungal), tumour

c) Alveolar shadowing

Fluffy, confluent shadows
= Airspace changes (alveoli, bronchioles)
Airspaces are filled up with:

Fluid	Pulmonary oedema
Pus	Pneumonia ("consolidation")
Blood	Goodpasture's

Look out for ***air bronchograms***: as alveoli fill with pus (white), bronchioles are outlined as they are still filled with air (black).

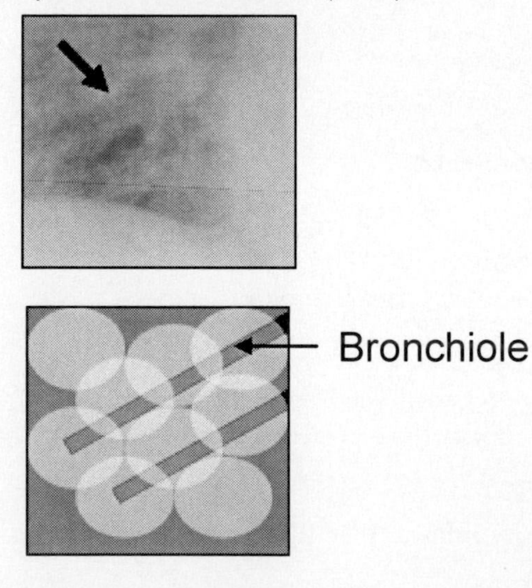

d) Reticular shadowing

Net-like shadowing

= Interstitial changes (surround bronchioles and vessels)

Fibrotic change/ fluid accumulation

Fibrosis, Sarcoidosis, alveolitis
May also be "reticulo-nodular" = net-like + small dots

e) Collapse

Confluent shadowing appears white like consolidation but there is *volume loss* and there are <u>no</u> bronchograms.

=proximal obstruction: neoplasm, mucus plug, foreign body

Loss of volume: mediastinal/tracheal shift (pulled)

Predictable locations. -Unfortunately beyond the scope of this book, for the keen, a good internet tutorial is available at:

http://www.radiology.co.uk/srsx/tutors/collapse/tutorial.htm#intro

10. Soft-tissues and bones

Present /symmetrical?

Look out for a mastectomy that makes a lung field look blacker
Look for fractures, especially if pneumothorax present.

11. Review areas

If you still cannot see any abnormality check review areas:

Apices

Lung peripheries

Beneath diaphragm

Behind heart

BLOOD GASES

Blood gas interpretation can be as easy or as difficult as you want it to be, for a simplistic yet workable method, try the following:

Look at pH:

Is it normal or abnormal?

Acidosis (<7.35) /alkalosis (>7.45)

Look at PCO_2:

Is it abnormal?

Does this explain or go against the pH?

With pH = Respiratory acidosis/alkalosis

Against pH = Metabolic acidosis/alkalosis
+ Respiratory compensation

To confirm your interpretation, look at HCO_3:

Is it abnormal?

Does this explain or go against the ph?

With pH = Metabolic acidosis/alkalosis

Against pH = Respiratory acidosis/alkalosis
+ Metabolic compensation

$\downarrow PO_2$, $\leftrightarrow/\uparrow PCO_2$	= Type I respiratory failure
$\downarrow PO_2$, $\uparrow PCO_2$	= Type II respiratory failure

Pathology

Chronic Obstructive Pulmonary Disease

Definition

A chronic, progressive respiratory disorder most commonly seen in smokers characterised by airway obstruction with little reversibility. COPD is an umbrella term that covers chronic bronchitis and emphysema.

Chronic bronchitis is a *clinical* diagnosis

= Cough productive of sputum

→ on most days for 3 months

→ for 2 consecutive years

Emphysema is an *anatomical* diagnosis

=Dilatation and destruction of airway spaces distal to terminal bronchioles. This can be determined on CT or histology.

Description

Incidence	Very common Smokers: 15% Non-smokers: 7%
Age	↑ w/age, > 35 year olds
Sex	M>F
Geography	Worldwide

Aetiology

Acquired	Smoking Pollution
Congenital	α1-antitrypsin deficiency

Pathophysiology

Chronic bronchitis (disease of the terminal bronchioles)

1. Hypertrophy of mucus glands + increased goblet cells

    → ↑ Sputum production

2. Inflammation: acute and chronic inflammatory cells

    → Airway narrowing

3. Squamous cell metaplasia + fibrosis

    → Airway narrowing

Emphysema (disease of the alveoli)

Inflammation results in damage to the alveolar walls. This leads to fewer but larger alveoli, with a decreased surface area.

↓ Lung recoil = ↓ Expiratory airflow = **air trapping**

↓ Surface area = ↓ Gas transfer

End result:

Type I respiratory failure	↓ PO_2 + ↔/↓ PCO_2
Type II respiratory failure	↓ PO_2 + ↑ PCO_2

Respiratory failure can eventually result in cor pulmonale: pulmonary vasoconstriction in response to hypoxia results in pulmonary hypertension. Right sided heart failure ensues when the right ventricle is unable to pump against the increased pressure.

Chronic CO_2 retention results in the respiratory centre desensitising to CO_2 levels and respiration is driven by hypoxia. This explains why giving high flow oxygen to COPD patients is dangerous –you take away their respiratory drive and they stop breathing!

History and Examination

Cause: Smoking (mcc)

Effect:

Symptoms	**Worsened by**
Cough + sputum	Cold
Wheeze	Pollution
Dyspnoea	Exercise

Medical Research Council Dyspnoea scale	
1	SOB on strenuous exercise
2	SOB walking up a slight hill
3	Stops for breath when walking at own pace
4	Stops for breath after 100m / few minutes on level ground
5	Breathless when dressing/ undressing

Signs

Observation	Cyanosis, tachypnoea, lip pursing
	Use of accessory muscles
Palpation	↓ Chest expansion
Percussion	Increased resonance due to hyperinflation
Auscultation	Quiet breath sounds, wheeze, expiration>inspiration

Signs of complications

Cor pulmonale	Raised JVP, bilateral leg oedema
Infection	Pyrexia, bronchial breathing/ crackles
Hypercapnoea	Peripheral vasodilation, bounding pulse, retention flap

Old fashioned terms:

Pink puffers = ↑ Alveolar ventilation (hence puffers)
= SOB but not cyanosed (hence pink)
= Type I respiratory failure

Blue bloaters = ↓ Alveolar ventilation
= Cyanotic but not SOB (hence blue)
= Type II respiratory failure

Investigations

Simple	**PEFR**	Airway obstruction / air trapping
	Sats	Hypoxia
	ECG	
	ABGs	(In acute setting)
		Type of respiratory failure, acidosis

Lab

Haem	**↑Hb/PCV**	Polycythaemia
	WBC	Infection
Chem	**CRP**	
	α1-antitrypsin	
Micro	**Sputum** mc&s	Infection

Imaging	Echo	Cor pulmonale/ cardiac function
	CXR	Hyperinflated lungs > 8 anterior ribs
		Flat diaphragm
		↓ Peripheral, ↑ proximal vessels
		CXR is often normal though

Special

Lung function tests:

These distinguish *obstructive* from *restrictive* lung diseases, and give an indication of the severity of either.

↓↓ FEV_1		
------------	= ↓ ratio	= Patient unable to expel air quickly
↓ FVC		= Obstructive disorder

FVC = Forced vital capacity, amount of air expelled from full inspiration to full expiration

FEV_1 = Percent of forced vital capacity can be expelled in the first second. This quantifies airflow obstruction:

Mild 50–80% predicted according to age/sex/height

Moderate 30–49% predicted

Severe < 30% predicted

The patient is able to breathe in but has difficulty *expelling* air due to air trapping and decreased lung recoil. When they breathe out, their airways collapse due to negative pressure. They take longer to breathe out (↓ FEV_1) and maintain their airways open with positive pressure obtained by pursing their lips.

Even though *total* lung capacity may be normal or increased due to air trapping and hyperinflation, their *vital capacity* is decreased.

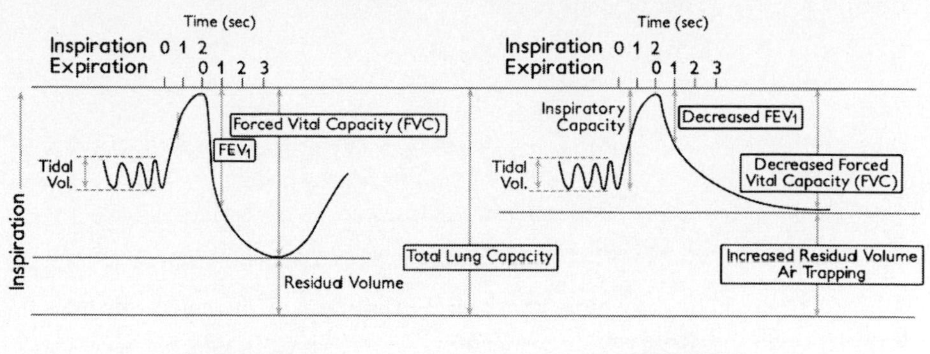

<u>Treatment</u>

Conservative

Smoking cessation

Physiotherapy

Medical

<u>Acute:</u>

ABC	
Oxygen	Requires careful administration through Venturi mask
	Recheck ABGs for CO_2 retention due to hypoxic drive
Antibiotics	Treat infective exacerbations aggressively
	(3rd gen. cephalosporin)
Bronchodilators	Ipratropium/ salbutamol via nebulisers
Corticosteroids	
Aminophylline	If no response to above
Diuretics	If cor pulmonale

<u>Chronic:</u>

Mild	Inhaled ipratropium ± inhaled salbutamol PRN
Moderate	Above plus:
	Salmeterol (long acting β2 agonist) ± inhaled steroid
Severe	Above ± oral steroid, nebulisers

Advanced Theophylline

Long term oxygen therapy -O_2 cylinder / concentrator
Indicated if PO_2 <7.3 or FEV_1 <1.5L
Available to **non-smokers only**
At least 15 hours per day to maintain PO_2 >8
Increases 3 year survival by 50%
Patients must not smoke as O_2 is flammable!

Interventional

BiPAP (Biphasic positive airway pressure) or intubation and mechanical ventilation in the acute situation. (***Not*** CPAP as the continuous positive pressure decreases their ability to breath out CO_2)

Bullectomy

Lung volume reduction surgery (increases elastic recoil)

Single lung transplantation (uncommon)

Prognosis

Variable. Morbidity is high. Mortality is falling, but if severely breathless, 5 year mortality is 50%

The only interventions that increase survival are smoking cessation and long term oxygen therapy.

ASTHMA

Definition

Recurrent episodes of dyspnoea, cough and wheeze due to reversible airway narrowing and thus obstruction, as a result of:

1. Hyper-responsive bronchial smooth muscle
2. Inflammation of the bronchi
3. Increased mucus production

Description

Incidence Very common

Age Intrinsic asthma: children
Extrinsic asthma: middle age

Geography Developed countries

Aetiology

Intrinsic asthma = no causative agent, *atopy*

Extrinsic asthma = definite external cause

Genetic factors

Family history

Association with atopy and other allergies: rhinitis, food allergies, anaphylaxis

Environmental stimuli

Specific	House dust mite, pollen, pets, dust, pollution Aspirin: increases leukotrienes
Occupational	Varnishes, paints, fumes, powders, animals, Plants, enzymes
Non-specific	Cold air, exercise, emotional stress

Pathophysiology

This is complex and not fully understood. There is chronic inflammation and bronchiole smooth muscle hypertrophy as a result of the release of a whole host of immune mediators such as IL3-5, IgE from B-lymphocytes and histamine and leukotrienes from Mast cells.

History and Examination

Cause: Family history, atopy, triggers including infective exacerbation.

Relationship to symptoms e.g. symptoms due to occupational triggers might improve on weekends.

Effect:

Symptoms

Episodic wheeze, cough, dyspnoea, chest tightness

Precipitants	Dust, cold, exercise, infective exacerbation
Diurnal variation	Worse early morning
Severity	Exercise tolerance
	Disturbed sleep/ nocturnal symptoms
	Frequent use of relieving medication
	Hospital/ITU admissions

Signs

Observation	Cyanosis, tachypnoea, use of accessory muscles
	Leaning forward, agitated
Palpation	↓ Chest expansion
Percussion	Hyper-resonant
Auscultation	Quiet breath sounds, polyphonic wheeze
	Length of expiration > inspiration
	Absence of wheeze is bad news if it means the patient is not moving enough air to cause it.

Also: Tachycardia

± **Pulsus paradoxus** = BP fall >5mmHg on inspiration

This is an *accentuation* of normal physiology

Severe attack: PEFR <50%

Speaking in incomplete sentences

Tachycardia >100bpm

Tachypnoea > 25 bpm

Life threatening attack: PEFR <33%

Silent chest

Exhaustion/ confusion

Bradycardia

Cyanosis

Rising CO_2

Investigations

Simple	**PEFR**	airway obstruction / air trapping
	Sats	
	Blood gases	(in acute setting) Type of respiratory failure, acidosis

	Mild	Moderate	Severe	ITU
PO_2	↑	↓	↓↓	↓↓↓↓
PCO_2	**N**	↓	**N/↑**	↑
pH	↑	**↑/N**	**N**	↓

Initially, the patient hyperventilates to increase PO_2, thus expelling CO_2. As severity increases, the patient becomes exhausted and CO_2 levels start to normalise and even rise, causing acidosis. Thus a normal but rising PCO_2 in an acute asthmatic is worrying.

Lab

Haem	**WBC**	Infection
Chem	**CRP**	
Micro	**Sputum** (mc&s)	Infection

Imaging

CXR Hyperinflated lungs > 8 anterior ribs

Flat diaphragm

CXR is often normal though

r/o pneumothorax

<u>Treatment</u>

Conservative

Patient education

Avoid allergens and triggers

Medical

<u>Acute:</u>

ABC

Sit patient up

Oxygen High flow oxygen through non-rebreather bag

Bronchodilators Salbutamol/ ipratropium via nebulisers

Corticosteroids	Prednisolone po or hydrocortisone i.v.
Fluids	Frequently dehydrated from hyperventilation
Antibiotics	Treat infective exacerbations aggressively

If continuing:

Back to back nebulised salbutamol

Consider iv magnesium

Aminophylline	No loading dose if patient is on theophylline
	Adjust dose according to P450 interactions
	↓ If patient on inhibitors
	(Cimetidine, cipro, erythro, COC, steroids)
	↑ If patient on inducers
	(Phenytoin carbamazepine, rifampicin)

If life threatening:

Inform ITU & seniors

Consider intubation and mechanical ventilation

Monitor effects of treatment via PEFR, Sats, ABGs

The patient may be discharged once PEFR >75% of normal

Chronic:

Patients should be on a combination of "reliever" medication such as a β2 agonist and a "preventer" such as an inhaled steroid according to severity.

Adapted from British Thoracic Society Guidelines 2005:

Step 1	β2 agonist bronchodilator
Step 2	Add inhaled steroid
Step 3	Add long acting β2 agonist
Step 4	Increase to high dose inhaled steroid
	Consider adding 4th drug: leukotriene antagonist, theophylline or modified release β2 agonist
Step 5	Add oral steroid and refer to asthma clinic

Remember to step down after 3-4 month stability

Inhalers

MDI: Metered Dose Inhaler (spray)

Easibreathe: Breath activated (spray)

Dry Powder Inhalers (DPIs):

Accuhaler	Looks like a UFO
Turbohaler	Looks like a rocket
Twisthaler	Looks like a rocket
Handihaler	Oval with flip lid

Methods of administration

Aerosol Metered dose inhaler ± spacer

Powder inhalers with powder in blisters

Nebulised

Theophylline

Theophilline has a narrow therapeutic window and should be monitored.

It inhibits phosphodiesterase (increasing cAMP) and reduces bronchoconstriction

Side effects include arrhythmias, seizures and GI upset.

Prognosis

Asthma is a chronic disease requiring maintenance treatment.

Risk factors for ↑ mortality include:

- Poor compliance
- ITU admissions
- Hospital admissions despite steroid treatment

Many patients who have suffered with childhood asthma tend to grow out of it. However, it may return.

Types of Inhalers

Generic	Brand name	Action	Colour	Inhaler	Comment
Salbutamol	Ventolin	B2 agonist	Blue	MDI, Accuhaler, Easibreathe	Asthma > COPD
Terbutaline	Bricanyl	B2 agonist	Light blue	Turbohaler	Asthma > COPD
Ipratropium bromide	Atrovent	Muscarinic antagonist	Grey with green cap	MDI	COPD > Asthma
Tioproprium bromide	Spiriva	Muscarinic antagonist	Grey oval	Handihaler	COPD > Asthma
Salbutamol + ipratriprium	Combivent	Mix	Grey	MDI	
Salmeterol	Serovent	Long acting B2 agonist	Green	MDI	Takes 30 mins to start working
Eformoterol	Oxis	Long acting B2 agonist	Blue bottomed rocket	Turbohaler	Works straight away
Beclomethasone	Becotide/ Becloforte/ Beclazone	Inhaled corticosteroid	Brown	MDI, Easibreathe	Doses of 50 to 250mg
Fluticasone	Flixotide	Inhaled corticosteroid	Orange	MDI, Accuhaler	Most potent steroid
Budesonide	Pulmicort	Inhaled corticosteroid	Brown bottomed rocket	Turbohaler	
Mometasone	Asmanex	Inhaled corticosteroid		Twisthaler	
Ciclesonide	Alvesco	Inhaled corticosteroid	White	MDI	Activated in lungs
Salmeterol + Fluticasone	Seretide	LABA + ICS	Purple	MDI, Accuhaler	COPD > Asthma
Budesonide + Eformoterol	Symbicort	LABA + ICS	Red bottomed rocket	Turbohaler	Asthma > COPD

COPD vs Asthma

	COPD	Asthma
Smoker / ex-smoker	Nearly all	Possibly
Symptoms under age 35	Rare	Common
Chronic productive cough	Common	Uncommon
Breathlessness	Persistent and progressive	Variable
Night time waking with breathlessness and/or wheeze	Uncommon	Common
Significant diurnal or day to day variability of symptoms	Uncommon	Common

OBSTRUCTIVE SLEEP APNOEA

Definition

Intermittent obstruction or collapse of the upper airway causing apnoeic episodes during sleep and terminated by partial arousal.

Description

Incidence Very common, under diagnosed

Sex M>F

Geography West, due to obesity

Aetiology

Obstructed airway

Obesity, especially large neck
Anatomically narrow airway
Large tonsils
Acromegaly, amyloidosis, hypothyroidism

Respiratory depressants may worsen symptoms

Alcohol
Sedatives

Pathophysiology

The airway is sucked closed during sleep when muscles are hypotonic. Patients partially awaken due to increased respiratory drive resulting from chest wall strain against a closed airway and hypoxia, allowing muscles to regain tone and reopen airway. The continuous wakening results in unrefreshing sleep.

History and Examination

CAUSE:

Obesity, airway obstruction

EFFECT:

Symptoms

Unrefreshing, restless sleep leading to daytime sleepiness
Snoring
Morning headache
↓ Performance at work, ↓libido

Signs

Obese patient with short, fat neck
Airway obstruction

Investigations

Polysomnography = sleep study
ECG + Sats + EMG
Diagnostic if >15 episodes of apnoea in one hour

Treatment

Conservative

Patient education
Weight loss
Avoid sedatives

Interventional

CPAP (Continuous positive airway pressure) at night
Tonsillectomy

Prognosis

Treatment, CPAP in particular, can vastly improve symptoms.

Bronchiectasis

Definition

Abnormal, permanent dilatation of airways due to repeated infection and chronic inflammation. Clinically characterised by the production of copious amounts of sputum.

Description

Incidence	Not uncommon, though prevalence falling
Age	Cystic fibrosis is mcc in children
Geography	Cystic fibrosis is mcc in developed countries

Aetiology

Risk factors:

Smoking
Recurrent infections
Mucocilary clearance defects: Cystic fibrosis, Kartagener's syndrome
Mechanical bronchial obstruction

Pathophysiology

There is a vicious cycle of infection causing tissue damage resulting in defective mucociliary clearance which in turn promotes infection.
Particular organisms include:

Pseudomonas
H. influenzae
Strep. Pneumonia
Staph. aureus

History and Examination

CAUSE: Family history, mucociliary defects, smoking

EFFECT:

Symptoms

Chronic cough
Copious purulent sputum production
Haemoptysis (may be massive)
Recurrent infections
Persistent halitosis

Signs

Observation	Clubbing, anaemia (of chronic disease), cyanosis
Auscultation	Coarse crackles, wheeze

Investigations

Simple	Sats	
	Blood gases	Severity
	Sweat test	CF
Lab		
Haem	Hb	Anaemia
	WBC	Infection
Chem	CRP	
Micro	Sputum mc&s	Infection

Imaging **CXR**

Tramlines = Thickened/mucus-filled bronchi
Consolidation from current infection
r/o pneumothorax

HRCT

Signet ring sign
= irregular thickening of the bronchial wall

Special	Lung function tests	Obstructive pattern

Treatment

Conservative

Physiotherapy Intensive postural drainage is very important
Smoking cessation

Medical

Prophylaxis Pneumovax (vaccination), long term antibiotics
Exacerbations Oxygen, prompt antibiotics
Bronchodilators if element of reversibility
Corticosteroids slow progression

Surgical

Excision of lung if localised disease, e.g. if caused by foreign body
Bilateral lung transplant

Prognosis

Complications of bronchiectasis include airway obstruction leading to respiratory failure, recurrent pneumonia, pneumothorax and pleural effusion.

Cystic Fibrosis

Definition

Multisystem autosomal recessive disease due to abnormal chloride secretion and sodium absorption. Hyperviscose secretions result in repeated respiratory infections and pancreatic insufficiency.

Description

Incidence	Most common autosomal recessive disease in Whites
	One in 25 Whites is a carrier
Age	Children
Geography	Whites>> Blacks >> Asians

Aetiology

Risk factor = family history, autosomal recessive inheritance

Pathophysiology

Abnormal cystic fibrosis transmembrane conductance regulator (CFTR) gene results in ↑ Cl secretion and Na absorption, which increases the viscosity of secretions. Effects are seen in the exocrine glands: lungs, pancreas, liver and reproductive tract.

History and Examination

CAUSE: Family history

EFFECT:

Respiratory	Bronchiectasis + clubbing Sinusitis + polyps Haemoptysis Pneumothorax Respiratory failure + cor pulmonale
Gastro	Meconium ileus Pancreatic failure Malabsorption + steatorrhoea Malnutrition Gallstones Diabetes
Reproductive	Infertility

Investigations

Simple	**Sweat test** BMs	
Lab		
Haem	Hb **WBC**	Anaemia Infection
Chem	CRP Glucose LFTs, amylase	 Diabetes Cirrhosis, pancreatitis
Micro	**Sputum** mc&s	
Imaging		
	CXR	Bronchiectasis Pneumonia Pneumothorax
	Abdo US	Cirrhosis, pancreatitis
Special	**PCR**	Genetic testing
	Lung function tests	Diagnosis Monitoring

Treatment

Involve multidisciplinary team

Conservative
Education
Genetic counseling
Dietician
Physiotherapy -intensive postural drainage is very important

Medical

Prophylaxis	Pneumovax, H. influenza immunization
	Long term antibiotics

Treat infections aggressively
Treat diabetes and complications
Pancreatic enzyme replacement

Surgical
Heart/lung transplantation
Gene therapy in the future

Prognosis

Prognosis has improved drastically in the last 40 years. Currently, adult patients tend to die in their 20-30's while children diagnosed today are given a life expectancy of around 40 years.

PNEUMONIA

Definition

Acute inflammation of the lower respiratory tract caused by infection.

Description

Incidence	Very common, community incidence - 2/1000 adults
	One in four require hospital admission
Age	Any
Sex	M:F

Aetiology

Risk factors

- Hospitalisation
- Smoking
- Aspiration e.g. stroke
- Underlying lung disease: bronchiectasis, COPD
- Immunosuppression
- IVDU

1. Community acquired pneumonia

Typical

Pneumococcus (mcc)	esp. Post viral infection
Haemophilus influenzae	esp. Underlying disease, COPD

Atypical

Chlamydia psittaci	Birds
Mycoplasma pneumoniae	esp. Young adults
Legionella	Water

2. Hospital acquired pneumonia (nosocomial)

Gram negative bacteria and anaerobes

E. coli

Pseudomonas	esp. Cystic fibrosis, ventilation
Klebsiella	esp. Alcoholics
Staph aureus*	esp. IVDU, v. severe

*also community acquired

3. Aspiration pneumonia

Oropharangeal and other gut anaerobes

4. Opportunistic

Pneumocystis carinii pneumonia (PCP)

Fungi: cryptococcus, aspergillus

Viruses: CMV, HSV

History and Examination

CAUSE: Risk factors as above, contacts

EFFECT:

Symptoms

Malaise

Dyspnoea

Cough + purulent sputum

Haemoptysis

Pleuritic chest pain

Signs

Fever

Confusion

Cyanosis

Tachypnoea

Tachycardia

Consolidation

Palpation	↓ Expansion, ↑ tactile fremitus
Percussion	Dull
Auscultation	Crackles, bronchial breathing, pleural rub, ↑ vocal resonance

Investigations

Aims of investigation are to confirm the diagnosis, define the cause, assess severity, identify complications and exclude cancer.

Simple	**Sats**	
	ABGs	Respiratory failure
	ECG	AF
Lab		
Haem	**WBC**	Infection (↑neutrophils)
Chem	**CRP**	
	U&E	Urea for severity, dehydration
Micro	**Sputum**	mc&s
Imaging	**CXR**	
	Consolidation = shadowing with air bronchogram	
	Pleural effusion = blunted costophrenic angle, meniscus	

Indications of severity "CURB 65" from BTS guidelines 2001

No factors present = may treat in the community

2 factors present = consider hospital admission

Confusion

Urea >7

Respiratory rate >30

Blood pressure: systolic<90 *or* diastolic <60

Age over **65** years

Treatment

Preventative

Pneumovax indications:

Heart/lung disease, cirrhosis, nephrosis, diabetes, immunosuppression

Supportive

Physiotherapy

Oxygen

Fluids

Analgesia

Ventilation

Curative

Antibiotics

Every hospital has its own antibiotic guidelines based on local prevalence of organisms and their sensitivities. These are regularly updated and you should familiarise yourself with them when you start at a new hospital.

Adapted from British Thoracic Guidelines 2001 (updated 2004)

Community acquired

Community Rx

Mild	**Amoxicillin**
Moderate	**Amoxicillin** plus **macrolide** po (e.g. erythro/clarithromycin to cover atypicals)

Hospital Rx

Moderate	**Ampicillin** iv plus **macrolide**
Severe	One of below iv plus **macrolide**

1. **Co-amoxiclav** (augmentin)
2. **2nd generation cephalosporin** (cefuroxime)
3. **3rd generation cephalosporin** (ceftriaxone)

Aspiration pneumonia

Cefuroxime + metronidazole

Hospital acquired/ immunosupressed

3rd generation cephalosporin (ceftriaxone)
± Gentamicin if sepsis/ failure to improve

Atypical

Legionella	Clarithromycin
Chlamydia	Tetracycline
Klebsiella	Cefuroxime
PCP	Trimethoprim
Fungi	-azoles/amphoteracin B

Prognosis

Complications:

Respiratory failure
Pleural effusion
Empyema
Lung abscess
Atrial fibrillation
Pericarditis/ myocarditis

Empyema = pus in pleural cavity
Recurrent fever, CXR shows pleural effusion
Chest aspirate/ drain yields turbid exudate with ↓ glucose, ↑LDH
Pus (or pH <7.2 if fluid clear) requires wide bore chest drain insertion.

Lung abscess = cavitating area in lung of localised, suppurative infection
Causes: lung infarction, pneumonia (esp. staph and klebsiella), aspiration
CXR shows cavitation with fluid level. Rx: draining/excision if no response to antibiotics.

Pleural Effusion

Definition

= Accumulation of fluid in pleural space. May be classified as transudate or exudate according to protein content.

Pathophysiology

Exudates <u>>30g/L of protein</u> (>200U/L LDH)

These are usually unilateral and result from infection, inflammation or malignancy. There is high protein content because **leaky capillaries** have allowed inflammatory cells and protein through.

Common	Bacterial pneumonia
↓	Carcinoma and infarction
	TB
↓	Connective tissue disease
	Post-MI
↓	Acute pancreatitis
Rare	Mesothelioma

Transudates <u><30g/L of protein</u> (<200U/L LDH)

These are usually bilateral and as a result of increased hydrostatic pressure or decreased oncotic pressure. There is low protein content because it is a result of **osmosis**. Capillary walls remain intact.

Increased hydrostatic pressure
Cardiac failure

Decreased oncotic pressure = Hypoalbuminaemia
Liver failure
Nephrotic syndrome
Burns
Malnutrition

For protein levels between 25-35, further analysis can be made using *Light's criteria.*

Exudates meet at least one of the following criteria, whereas transudates meet none, (blood must be taken at the same time as pleural fluid):

1. The ratio of pleural fluid protein to serum protein is greater than 0.5
2. The ratio of pleural fluid LDH and serum LDH is greater than 0.6
3. Pleural fluid LDH is more than 2/3 normal upper limit for serum

History and Examination

CAUSE: Look for features of underlying cause, screen for:

Heart failure	Orthopnoea, ↑ JVP, oedema, bibasal creps
Malignancy	Constitutional symptoms, haemoptysis
Infection	Fever, purulent sputum

EFFECT:

History	SOB, cough
Palpation	↓ Expansion, ↓ tactile fremitus
Percussion	Stony dull
Auscultation	↓ Breath sounds, ↓ vocal resonance

Investigations

Simple	ECG, ABGs	
Lab		
Haem	**FBC**	Infection
	Clotting	Liver function
Chem	**CRP**	Infection
	LFTs	Liver failure, protein
		Tumour markers
Imaging	**CXR**	Blunted costophrenic angle
		Meniscus
	US	Confirm diagnosis
		Guide pleural tap
Invasive	**Pleurocentesis**	Diagnostic ± therapeutic.

The appearance of pleural fluid can give clues as to the cause:

Clear	Transudate	Pulmonary oedema
Serous	Exudate	TB, RA
Cloudy	Exudate	Infection
Bloody	Exudate	CA, PE, TB

Pleural fluid should be sent for analysis:

Chemistry	Protein, glucose, pH, LDH
Cytology	Malignancy
Immunology	RF, ANA, complement
Microbiology	Mc&s, ± TB culture and ZN stain

Treatment

Investigate for and treat underlying cause
Drainage: pleural aspiration if small or chest drain if larger
Pleurodesis with talc or tetracycline/ bleomycin if recurrent

Prognosis

Prognosis is dependent on the cause.

PNEUMOTHORAX

Definition

= Air in pleural space

Description

Incidence	Common
Sex	M>F

Aetiology

1° **Spontaneous**
Tall, thin men with family history

2° **Underlying lung disease**
Asthma/COPD/TB/pneumonia/bronchiectasis

Trauma
Penetrating injury e.g. stab wounds, broken ribs

Iatrogenic
CVP line, positive pressure ventilation, biopsies, pleural taps

Pathophysiology

Open

Communication between pleural space and exterior, due to a chest wall defect, e.g. stab wound

Closed

Communication between pleural space and airways

Tension ***medical emergency***

A tension pneumothorax may develop whether open or closed and is dependent on the formation of a one-way valve between the communication. When negative pressure is created on inspiration, air is sucked into the pleural space. However due to the valve, the air is prevented from leaving on expiration. The accumulating air that is unable to escape increases intrathoracic pressure, decreasing venous return and cardiac output.

History and Examination

CAUSE:	Underlying lung disease, trauma
EFFECT:	
History	Sudden onset SOB + pleuritic chest pain/ asymptomatic
Palpation	↓ Chest expansion on the affected side
Percussion	Hyperresonant
Auscultation	↓ Breath sounds
Tension	*Tracheal deviation*
	Shock: tachycardia, hypotension → cardiac arrest

Investigations

Treat immediately if suspected tension pneumothorax
–don't wait for the CXR.

Simple	**ABGs**	
Lab		
Haem	FBC	Infection
Chem	CRP	Infection
Imaging	**CXR**	Pleural edge
		No lung markings beyond pleural edge
		Underlying cause
	CT	May pick up small pneumothorax

Treatment

Adapted from British Thoracic Society Guidelines 2003

Treatment depends on symptoms and extent of pneumothorax on CXR

Conservative
If asymptomatic, spontaneous pneumothorax with <2cm rim of air on CXR may need no treatment other than advice.

Aspiration
1. For a primary pneumothorax patient who is symptomatic and/or those with >2cm rim of air on CXR

2. All secondary pneumothoraces
If >50 years, SOB and >2cm rim of air on CXR, go straight for a chest drain

Chest drain
If aspiration was unsuccessful, the next step is an intercostal drain.
Primary treatment if >50 years, SOB and >2cm rim of air on CXR

Tension pneumothorax

Insert large venflon into 2nd intercostal space, midclavicular line as soon as suspected. This is then followed by a chest drain.

Prognosis

Prognosis is good. Recurrence rates for spontaneous pneumothorax are 30% in 5 years.

PULMONARY EMBOLUS

Definition

= Thrombus most commonly from systemic veins dislodges and embolises via the right heart into the pulmonary arteries

Description

Incidence Very common, often sub-clinical

Sex F>M

Aetiology

PE first requires thrombus formation, usually in the deep veins of the leg (DVT). Other causes of emboli are rare and include post-MI right heart thrombus, septic emboli (endocarditis), neoplastic cells, fat/ air/ amniotic fluid.

Virchow's triad for thrombus formation

Stasis	*Immobility*	
		Recent surgery, esp. orthopaedic
		Prolonged bed rest
	↑ *Abdominal pressure*	
		Late pregnancy
		Pelvic surgery
		Pelvic neoplasm
Damaged wall	Trauma, inflammation	
Hypercoagulability	Malignancy	
	Pregnancy, COC, HRT	
	Factor V Leiden, protein C/S deficiencies	

Previous history of a venothrombotic event (VTE=DVT/PE) is highly significant. Smoking is no longer thought to be a risk factor.

Pathophysiology

The embolus leads to V/Q mismatch as the affected part of the lung is ventilated but not perfused. Alveoli may collapse and eventually infarct. A large embolus may cause pulmonary hypertension and cor pulmonale. Massive PE is a very common cause of unexpected, natural sudden death.

History and Examination

CAUSE:

Wells' DVT criteria 2003: Score one point for each of the following:

- Active cancer
- Paralysis/ immobilization
- Bedridden >3days / major surgery in last 12 weeks
- Localised tenderness along distribution of deep veins
  - (Behind knee then medially to groin)
- Entire leg swollen
- Calf swelling discrepancy >3cm
- Pitting oedema (greater in symptomatic leg)
- Previously documented VTE

If an alternative diagnosis is more likely than DVT, subtract 2 points

-2 to 0	Low risk, probability of DVT = 3%
1 to 2	Moderate risk, probability of DVT = 17%
>2	High risk, probability of DVT = 75%

Effect:

May be asymptomatic (60-80% of patients with DVT develop PEs but only a small percentage know about it)

There are 3 syndromes:

1. *Pleuritic chest pain + haemoptysis* (small to moderate PE)
Occurs as a result of pulmonary infarction

2. *Dyspnoea + hypoxia* (moderate to large PE)
Pleural rub, coarse crackles ± pleural effusion

3. *Circulatory collapse* (large to massive PE)
Sudden collapse due to acute obstruction of right ventricle outflow
Severe central chest pain, pale, clammy
↑ JVP, right parasternal heave (Pulmonary hypertension)
Gallop rhythm, widely split heart sounds
(Pulmonary valve closing later than aortic valve)
Shock→ cardiac arrest→ death

Investigations

Aims are to diagnose/exclude PE and diagnose underlying cause, especially *malignancy.*

Initially the patient is scored according to Well's criteria and a D-dimer is requested. A positive D-dimer or moderate to high Well's score warrants further investigation. Those who have a low score with negative D-dimers are considered to have PE excluded.

Due to its high negative predictive value, D-dimer is more useful in ruling out PE than diagnosing it. Other conditions such as pregnancy, malignancy and inflammation raise D-dimer levels, so a positive result does not imply PE. However, almost all patients with DVTs have positive D-dimers.

Wells' PE criteria 2000

Suspected DVT	3 points
PE most likely diagnosis	3 points
Heart rate >100bpm	1.5 points
Immobilisation/ surgery in last 4/52	1.5 points
Previous VTE	1.5 points
Haemoptysis	1 point
Known active malignancy	1 point

<2 points	Low, probability of PE = 4 %
2 to 6 points	Moderate, probability of PE = 21%
>6 points	High, probability of PE = 67%

Simple	**ABGs, sats**	
	ECG	**S_1/Q_3T_3**
		Classic but uncommon: S wave in lead I Q wave and inverted T in lead III **Cor pulmonale** RAD, peaked P waves, RBBB
Lab	**D-dimers**	**See above**
	FBC	Infection, malignancy
	LFT	Malignancy
	WBC/ESR	↑ may indicate infarction
Imaging	CXR	Most likely normal
		Excludes other conditions
	US leg	DVT
	Echo	Large thrombi/ RV strain (massive PE)
	VQ scan	Fails to detect 50%
	CTPA	=CT pulmonary angiogram
		Diagnostic investigation of choice
	MRA	=Magnetic resonance angiography
		High sensitivity and specificity
	Angiography	Gold standard but invasive

Treatment

Prophylaxis

Thromboembolic deterrent (TED) stockings
Anticoagulation with warfarin/ heparin to prevent recurrence
LMW heparin sc for all immobile patients
Mobilisation
Stop COC/HRT prior to surgery

Medical

ABC, massive PE may cause cardiac arrest!
Oxygen
Thrombolysis (massive PE)

Anticoagulation (this is prophylactic –it won't dissolve clot that has already formed)

LMW heparin sc (e.g. Tinzaparin, Enoxaparin)

For high risk patients awaiting imaging
Until INR in therapeutic range

Warfarin once PE confirmed, INR target 2-3

3 months if first time, idiopathic
6 months if other cause
Lifelong treatment for recurrent VTEs
Isolated above knee DVTs should be treated for 6 months

Interventional

Embolectomy (massive PE)
Inferior vena cava filter (recurrent PE)

Prognosis

10% mortality, otherwise good prognosis.

TUBERCULOSIS

Definition

The most common infective cause of death worldwide, TB is a granulomatous disease caused by mycobacterium tuberculosis affecting any part of the body except hair and nails. It is a notifiable disease.

Description

Incidence	Extremely common worldwide, prevalence 2 billion
Age	Any
	More prevalent in children/elderly as ↓ immune system
	Highest mortality in young adults
Geography	Asian subcontinent, West Indies, Africa

Aetiology

Mycobacterium tuberculosis: an acid fast bacillus, spread by respiratory droplets.

Risk factors:

Contacts
Travel
Immunosuppression: HIV, malignancy, diabetes, steroids

Pathology

Any part of the body may be affected

Primary infection– *small focus in the tissue but extensive involvement of the draining lymph nodes*

Most commonly pulmonary, may be GI (ileo-caecal junction)

Formation of Ghon focus = initial area of infection

Followed by Ghon complex = draining lymph nodes plus lung lesions

These provide a hiding place for the mycobacterium away from the immune system in caesiating granulomas (see sarcoidosis below) which eventually fibrose and calcify.

Post-primary/ Secondary TB– *large focus in the tissues with only small lymph node response*

Reactivation can occur with any form of immunocompromise.

Pulmonary

Primary pulmonary = cavitation in apex
Miliary = indicates haematogenous spread

Meningeal

Genitourinary

Bone (Pott's disease)

Skin (Lupus vulgaris)

Adrenal (Addison's disease)

Pericarditis

Peritoneal

History and Examination

CAUSE: Contacts, immunosuppression

EFFECT:

TB can manifest itself anywhere in the body and can therefore be considered as a differential for almost any symptom. Investigation is based on isolating acid fast bacilli

Pulmonary TB (mc)

Effects range from asymptomatic to:
Weight loss, malaise
Cough, haemoptysis
Fever, night sweats, lymphadenopathy

Ix: *CXR:* Apical cavitation, consolidation, fibrosis or calcification ± pleural effusion

Miliary TB has ground glass appearance (multiple diffuse nodules)

Sputum: 3x samples for mc&s staining for acid fast bacilli

Meningeal TB

Subacute: fever, headache, N&V, neck stiffness, photophobia
Focal signs suggest granuloma

Ix: *LP:* ↓ glucose, ↑↑ protein, lymphocytes, no organisms

Genitourinary TB

Sterile pyuria, dysuria, haematuria, back pain

Ix: *MSU:* 3 early morning urine samples: acid fast bacilli

Renal US

Bone TB

Collapse and angulation of the spine, back pain, ± paravertebral abscess

Ix: *Spine Xray, biopsy*

MRI: r/o cord compression if neurology

Peritoneal TB

Abdo pain, GI upset, ascites
Ix: *Ascitic tap:* acid fast bacilli
Laparotomy

TB pericarditis

Acute pericarditis ± effusion
Chronic pericardial effusion/ Chronic constrictive pericarditis (due to granulomata)

Ix: *Echocardiography*

Skin

Lupus vulgaris = nodules on neck/face

Ix: *Biopsy*

Investigations

Aims of investigation are to confirm the diagnosis by demonstrating the presence of mycobacterium tuberculosis. Site-dependent investigations are in italics above.

MC&S

Definitive diagnosis is only made by culturing the mycobacterium, but this can take up to 8 weeks. Faster answers can be obtained using the **Ziehl-Neelsen stain** for acid fast bacilli or TB PCR.

Histology

May show caeseating granuloma. Only TB/fungi form *caeseating* granulomas, i.e. Crohn's granulomas do *not* caesiate.

Immunology

Mantoux / PPD (purified protein derivative/tuberculin)

This is a Type IV (delayed) hypersensitivity reaction, so results are read 2-3 days later. Following intradermal tuberculin, redness and induration >10 mm in diameter implies previous exposure to tuberculin, i.e. anything from previous immunisation to active disease. A strong reaction may indicate active TB. The reaction may be weaker in immunosuppression.
The Heaf test, used for large-scale screening, is similar but less accurate.

Treatment

Preventative

BCG (bacille Calmette-Guérin)

School vaccination (at around 13 years)
Given only to Mantoux/Heaf negative children
Decreases risk of TB by 70%

Contact tracing
Family, close friends and those sharing living space with patients are given chemoprophylaxis: isoniazid for 6 months.

Supportive

Physiotherapy
Oxygen
Fluids
Analgesia

Curative

Antibiotic treatment is based on combination therapy of 2-4 drugs, depending on culture sensitivities. Clinical response supports diagnosis.

Pulmonary TB: Short course chemotherapy x 6 months
Bone, meningitis: Long course chemotherapy x 12 months

NICE 2006 guidelines for pulmonary TB are 6 months of isoniazid and rifampicin initially, plus pyrazinamide and ethambutol for the first 2 months. Remember this as "4 drugs for 2 months, then 2 drugs for 4 months".

Blind treatment of TB (**RIPE**)	**Side effects**
Rifampicin	Hepatitis
Isoniazid	Hepatitis, neuropathy
Pyrazinamide	Hepatitis
Ethambutol	Optic neuritis (reversible)

Due to the possible side effects, LFTs, U&E, FBC, colour vision and acuity should be checked before and during treatment. Steroids should be added in meningeal, GU and especially pericardial disease. Surgery may be considered in constrictive pericarditis and Pott's disease.

Prognosis

Complications:

Cavitation leading to aspergilloma
Pleural effusion
TB empyema
Massive haemoptysis

SARCOIDOSIS

Definition

A multisystem granulomatous disease of unknown cause.

Description

Incidence	Not uncommon
Age	peaks at 20-40 years
Sex	F>M (slight)
Geography	Great geographical variation Common in USA, uncommon in Japan

Pathophysiology

Any organ (most commonly lung, eyes, skin) may be affected by the formation of granulomas. Like TB, think of this disease in any differential diagnosis!

Granulomas: Nodular growths of epithelioid macrophages (± lymphocytes, multinucleated giant cells, eosinophils and plasma cells) that are walled off by an area of fibrosis. Note the only granulomas that caesiate are those caused by TB or fungus.

In sarcoid, granulomas are non-caesiating and may secrete ACE and 1α-hydroxylase (→ vitamin D) thus causing hypercalcaemia.

Acute	Erythema nodosum, arthralgia, lymphadenopathy (good prognosis with steroids)
Chronic	Slowly progressive lung fibrosis (poorer prognosis)

History and Examination

Asymptomatic

Following routine CXR showing bilateral hilar lymphadenopathy

Pulmonary (mc)

Cough, progressive dyspnoea, chest pain and lymphadenopathy

Non-pulmonary

Lymphadenopathy	
Skin	Erythema nodosum, lupus pernio, nodules
Occular	Uveitis, conjunctivitis
Bones	Arthritis (small joints), bone cysts
Liver	Hepatosplenomegaly
Brain	Uncommon, but severe: focal signs
Heart	Rare: heart block/ failure

<u>Investigations</u>

Simple		ECG, ophthalmology assessment (slit lamp)	
Lab	Haem	WBC	Lymphopenia
		Platelets	Thrombocytopenia
	Chem	**ESR**	Raised
		Calcium	Raised
		ACE	Useful for monitoring disease
Imaging			
		CXR	Abnormal in 90%
		HRCT	

Stage 0	normal
Stage 1	bilateral hilar lymphadenopathy (BHL)
Stage 2	BHL + pulmonary infiltrates
Stage 3	pulmonary infiltrates only
Stage 4	pulmonary fibrosis/pleural involvement

Special	**Lung function tests**	Restrictive picture
Invasive	Biopsy showing non-caesiating granuloma (diagnostic)	

Treatment

Conservative

Monitor asymptomatic patients with bilateral hilar lymphadenopathy

Medical

Acutely:

Bed rest, NSAIDs for arthralgia

If parenchymal lung disease, uveitis, hypercalcaemia or neuro/cardiac involvement:

Steroids

If severe:

Cyclosporine, cyclophosphomide

Treat hypercalcaemia as appropriate

Prognosis

More severe and higher mortality in Blacks, elderly or diffuse disease.

60% with thoracic disease only spontaneously recover in 2 years.

Cryptogenic Fibrosing Alveolitis

Definition

Idiopathic inflammatory disease of the alveoli which results in pulmonary fibrosis (restrictive lung disease).

Also known as "*intrinsic* allergic alveolitis" as there is no known external antigen.

Description

Incidence	Relatively rare
Age	late to middle age
Sex	M>F

Pathophysiology

Complex pathophysiology. Macrophages release growth factors resulting in collagen deposition. Histologically the cellular infiltration causes thickening and fibrosis of alveolar walls and alveolitis (immune cells in alveolar space).

Fibrosis (scarring) causes a restrictive picture with reduced compliance of the lungs, and increased lung recoil. Stiff lungs result in reduced lung volumes. Patients with restricted lung disease find it difficult to *breathe in* whereas patients with obstructive lung disease suffer from air trapping and find it difficult to *breath out*. Both result in smaller functional lung volumes.

History and Examination

Symptoms	Progressive dyspnoea and cough
	Arthralgia, malaise, weightless
Signs	Cyanosis
	Clubbing
	Decreased chest expansion
	End-inspiratory creps

Complications include pneumothorax, respiratory failure and pulmonary hypertension

Investigations

Simple	ABG	Respiratory failure
Lab		
Haem	WBC	Superimposed infection
Chem	**ANA**	30% positive
	RF	10% positive
Imaging	**CXR**	Diffuse ground glass shadowing
		Reticular-nodular changes
		Honeycombing
	HRCT	Above changes better visualised
Special	**Lung function tests**	Fibrosis =restrictive picture
Invasive	Biopsy/ bronchioalveolar lavage	

Treatment

Medical	Long term oxygen therapy
	Prednisolone
	Azathioprine
Surgical	Single lung transplant

Prognosis

Poor, mean survival is 5 years, with high mortality if acute. Lymphocytes on biopsy have a better prognosis than one characterised by neutrophil infiltration.

INDUSTRIAL LUNG DISEASE

Many lung diseases are a result of industrial exposure (i.e. at work) to factors such as dusts, gases, fumes and vapours. Asthma and fibrosis are the most common resulting diseases and their character depends on the dust inhaled.

Organic	→Extrinsic allergic alveolitis
Inorganic	
1. Fibrogenic	→ Pneumoconioses: Asbestois, silicosis, Coal workers'
2. Nonfibrogenic	→Iron, tin, chromium

Extrinsic Allergic Alveolitis

Definition

Hypersensitivity to inhaled *organic* dusts leading to inflammation and fibrosis of distal airspaces.

Description

Incidence	Uncommon
Age	Middle age

Aetiology

A variety of fungal, bacterial and animal proteins have been implicated.

Mouldy hay	Farmer's lung
Birds	Bird fancier's lung
Barley	Maltworker's lung
Others	Occupations with sawdust, cotton, cork etc.

Pathophysiology

Granulomatous reaction to organic dusts. Inflammation of terminal bronchioles extends distally into alveoli and there is progressive fibrosis.

History and Examination

Acute	4-8 hours post exposure SOB, cough Fever, myalgia, headache
Chronic	Progressive, exertional dyspnoea History of acute episodes Complications: respiratory failure/ cor pulmonale

The inspiratory crackles of fibrosis are heard on auscultation.

Investigations

Simple	ABG	Respiratory failure
Lab	WBC	↑ Neutrophils
Imaging	**CXR/HRCT**	Diffuse ground glass shadowing
		Reticular-nodular changes
Special	**Lung function tests** (restrictive picture)	
Invasive	Biopsy/ bronchioalveolar lavage	

Treatment

Conservative	Avoid/ reduce exposure e.g masks, protective clothing
	Compensation via UK Industrial Injuries Act
Medical	Oxygen
	Prednisolone
Surgical	Single lung transplant

Prognosis

Poor, mean survival is 5 years, with high mortality if acute. Lymphocytes on biopsy have a better prognosis than neutrophils.

Fibrogenic Pneumoconioses

All the following *inorganic* materials cause fibrosis, giving a similar picture to extrinsic allergic alveolitis

Asbestosis

Small asbestos fibres penetrate distally where they are engulfed by macrophages, setting off an inflammatory reaction culminating in fibrosis.

M>F

Symptoms occur decades after exposure, smoking is synergistic.

CXR may also show pleural plaques as well as fibrosis.

Complications include ***mesothelioma***, a malignant growth of the pleura. This can be accompanied by pleural effusion yielding bloody fluid on pleurocentesis.

Coal worker's pneumoconiosis

This was previously common in coal workers due to inhalation of coal dust - now rare due to the decline in the mining industry and the increasing demands made by Occupational Health departments.

Simple	Small nodules on CXR, asymptomatic
Complicated	Large nodules, cavitation, emphysema, fibrosis
Caplan's Syndrome	Pneumoconiosis +RA +pulmonary RA nodules

Silicosis

Now rare, seen after exposure to silica found in metal mining, stone quarries and ceramics.

CXR shows fibrosis and egg shell calcification of hilar nodes.

It has a good prognosis if chronic, provided exposure stops.

Acute disease can cause death in months as it is due to high levels of exposure.

Non-fibrogenic Pneumoconioses

This is a result of exposure to metals such as iron, tin and chromium. Although CXR changes are present, it is commonly asymptomatic.

RESPIRATORY FAILURE

Definition

Inadequate gas exchange resulting in hypoxia, PaO_2 < 8kPa, subdivided into 2 types according to PCO_2.

Pathophysiology

Note that many of the diseases that cause respiratory failure cause both decreased oxygenation (Type I) and abnormal ventilation (Type II).

Type I =↓oxygenation **Normal or low CO_2**

PCO_2 may be low as the patient hyperventilates to increase oxygenation, thus blowing off CO_2. Causes include:

↓ *Inspired O_2*	High altitude
Diffusion impairment	Pneumonia, pulm oedema, fibrosis, ARDS
VQ mismatch	PE, obstructive*/restrictive lung disease
R to L shunt	ASD/VSD/PDA

Type II= ventilatory failure **High CO_2**

Pulmonary disease	Asthma, COPD*, fibrosis
↓*Respiratory drive*	Sedative drugs, CNS tumour, trauma
Neuromuscular disease	Myasthenia gravis, Guillan-Barre, ALS, polio
Thoracic wall disease	Flail chest, kyphoscoliosis

*Note then that COPD can lead to either.

History and Examination

CAUSE: Features of underlying cause

EFFECT:

Hypoxia

Dyspnoea, central cyanosis
Restlessness, agitation, confusion
Cool, clammy, tachycardia
Longstanding: polycythaemia, pulmonary HTN, cor pulmonale

Severity can be clinically assessed by:

Use of accessory muscles
Pulsus paradoxus
Incomplete sentences
Asynchronous respiration
-rate of thoracoabdominal movement
Paradoxical respiration
-direction of thoracoabdominal movement

Hypercapnia

Headache
Confusion, drowsiness, coma
Warm, vasodilation, tachycardia
Bounding pulse, CO_2 retention flap, papilloedema

Investigations

The aim is to determine the type of respiratory failure and its severity as well as identify the underlying cause.

Simple	**ABGs**		
		Type I	**↓PO_2 / normal or ↓PCO_2**
		Type II	**↓PO_2, ↑PCO_2**
	ECG		Cor pulmonale
Lab	FBC		Infection, polycythaemia
	CRP		Infection
Imaging	**CXR**		Underlying cause

Treatment

1. Treat underlying cause

2. Treat hypoxia (↑ oxygenation)

Low concentration oxygen

Hudson mask "normal" mask through which you are unable to control the exact amount of oxygen delivered.

Non-rebreathing "high flow" mask delivers the highest concentration of oxygen of any of the masks (also uncontrolled).

Venturi mask delivers a fixed percent of oxygen 24%, 28% or 40%. Useful in COPD patients with hypoxic drive.

High concentration oxygen has to be given via a closed system

Bag and mask (tight fitting)
Endotracheal intubation

3. Treat hypercapnia (↑ ventilation)

↑ Respiratory drive: reverse sedatives, give stimulants
↑ Ventilation: Non-invasive/invasive

Although methods of ventilation have seemingly confusing names, if you understand the nomenclature, you can understand exactly what each method is providing.

Positive pressure means air is pushed into the lungs. It can be set to occur at different times during the respiratory cycle, i.e. on inspiration (to push air in) or expiration (preventing airway collapse whilst permitting a pressure drop for lung recoil). It can also be set at different levels (CPAP vs. BIPAP) and may or may not be synchronised with the patient's own breathing. Remember that expiration occurs as a result of the lung's natural recoil.

NIPPV: *Non-invasive* positive pressure ventilation

1. CPAP: Continuous Positive Airway Pressure

Constant positive pressure is provided, i.e. air is continuously pushed into the lungs, aiding inspiration and maintaining the airway open in expiration. Available as invasive and non-invasive. Used in obstructive sleep apnoea and pulmonary oedema. Best avoided in COPD as may exacerbate air trapping.

2. BIPAP: Biphasic Intermittent Positive Airway Pressure

As above but switches to a lower positive pressure on expiration. This means the patient does not have to fight against high positive airway pressure to breath out whilst the small amount of positive pressure continues to maintain airways open. Useful in COPD patients.

IPPV, *Intermittent* positive pressure ventilation

Positive pressure pushes air into the lungs causing inspiration. This pressure is then withheld (hence "intermittent"), causing expiration due to the lung's natural recoil. This can be synchronised with the patient's own respiratory effort as SIMV or "synchronised intermittent mandatory ventilation".

PEEP, Positive end expiratory pressure

This can be added to IPPV and SIMV to maintain the airways from collapsing at the end of expiration i.e. it is a small boost just at the end of expiration.

Ventilators provide either a fixed *tidal volume* (which may cause a pneumothorax if the lungs aren't compliant, requiring high pressures to blow in the set volume), or fixed *pressures*, which is dependent on the compliance of the lungs but may not necessarily shift adequate volumes.

Prognosis

Complications of mechanical ventilation:

Immediate

Trauma to anatomical structures including teeth and vocal cords
Misplaced tube: down one bronchus/ oesophagus

Early

Leak around tube
Displacement of tube
Obstruction
Pneumothorax

Late

Mucosal oedema/ulceration
Damage to cartilage
Fibrosis and tracheal narrowing

ACUTE RESPIRATORY DISTRESS SYNDROME

Definition

= Respiratory failure complicating pulmonary or systemic conditions that activate inflammatory mediators and result in:

Diffuse lung injury
Damaged pulmonary vasculature
Non-cardiogenic pulmonary oedema

Aetiology

Systemic	Sepsis, especially septic shock Hypotension Massive blood transfusion Disseminated intravascular coagulation (DIC) Drugs (illicit/ prescribed) Obstetric: pre-eclampsia, amniotic fluid embolism Burns Acute pancreatitis Head injury / ↑ ICP
Pulmonary	Pneumonia Aspiration Vasculitis Lung contusion

Pathophysiology

ARDS is pulmonary oedema as a result of 'leaky' inflamed lungs. It is the earliest manifestation of a generalised inflammatory response and multi-organ dysfunction syndrome.

Non-cardiogenic pulmonary oedema is the cardinal feature. Other pathological features include:

Pulmonary hypertension
Haemorrhagic intra-alveolar exudate
Hyaline membrane formation
Fibrosis (develops slowly, seen weeks later)

History and Examination

CAUSE: Underlying cause as above

EFFECT: Patient who is already unwell developing sudden onset tachypnoea, dyspnoea and hypoxia. Fine crackles of pulmonary oedema are heard on auscultation.

Investigations

Simple	**ABG**	Respiratory failure
Lab		
Haem	FBC Clotting	Infection DIC

Chem	U&E CRP Amylase	 Pancreatitis
Micro	Blood mc&s	
Imaging	**CXR**	Bilateral pulmonary infiltrates Ill-defined shadowing, esp. peripheral Normal sized heart Clues to underlying cause
Invasive	Pulmonary capillary wedge pressure <18mmHg	

Treatment

ITU admission + monitoring
Supportive treatment: Cardio-respiratory and nutritional
Treat underlying cause, esp. sepsis
In addition:

> Diuretics/ fluid restriction: ↓ volume overload
> Nitric oxide: ↓ V/Q mismatch and pulmonary hypertension
> High dose steroids: in late stages, controversial
> Surfactant: its role is unclear, though helpful in neonatal RDS.

Prognosis

Mortality is high at around 50% though dependent on cause.
Poor prognostic factors include increasing age and multi-organ failure.

LUNG CANCER

Definition

= Malignant tumour of the lungs. The most common cancer in the West and third most common cause of death in the UK after heart disease and pneumonia.

Description

Incidence	Very common
Age	↑ with age
Sex	M>F
Geography	Worldwide but important cause of mortality in the UK

Aetiology

As with many cancers, aetiology is a combination of genetic predisposition and environmental triggers.

Environmental risk factors:

Smoking
Asbestos, arsenic, chromium
Radiation
Cryptogenic fibrosing alveolitis (or indeed any fibrosis)

The increased risk in smokers is proportional to the number of cigarettes they smoke a day, crudely: 5 cigarettes a day = 5x the risk of lung cancer. By giving up, the patient will roughly decrease their risk by half every 5 years.

Pathophysiology

Small cell carcinoma (20%)

Central, rapid growth
Endocrine cells which secrete ADH and ACTH
Responds to chemotherapy but has poor prognosis

Non small cell

Squamous (40%)

Central, large airways
May cavitate, obstruct the airway and predispose to infection
Secretes PTH causing hypercalcaemia
Slow growth and late metastasis

Adenocarcinoma (20%)

Peripheral lung
Most common form in non-smokers, F>M, the elderly and following asbestos exposure
Slow growth and late metastasis (but earlier than squamous)

Large cell (10%)

Less well-differentiated tumour with early metastasis

Alveolar cell (<1%)

History and Examination

CAUSE: Risk factors as above.

EFFECT: Any patient with haemoptysis or unexplained symptoms (listed below) lasting >3 weeks should have an urgent CXR as recommended by NICE guidelines.

For any cancer, classify the possible effects into those of the tumour itself, constitutional, local invasion, metastases and paraneoplastic syndromes.

Tumour

Cough, SOB, haemoptysis
Complications: infection, collapse, pleural effusion

Constitutional

Weight loss, malaise, cachexia, clubbing, and anaemia

Local invasion

Recurrent laryngeal nerve palsy	Hoarseness
Horner's syndrome	Miosis, ptosis, anhydrosis
Brachial plexus involvement	C8, T1, T2: severe pain, Wasted small muscles of the hand
SVC obstruction	Facial oedema, morning headache, ↑JVP
Phrenic nerve palsy	Paralysis of hemidiaphragm, SOB
Oesophagus	Dysphagia
Heart	Arrhythmias, pericardial effusion
Chest wall	Pain

Metastasis

Bone	Pain, hypercalcaemia, fractures
Liver	Asymptomatic/pain, deranged LFTs, ascites
Brain	Headache, confusion, focal signs
Adrenals	Addison's disease

Paraneoplastic

SIADH	*Small cell carcinoma* →ADH
Cushing's disease	*Small cell carcinoma* → ACTH
Hypercalcaemia	*Squamous cell carcinoma* →PTH

Eaton-Lambert Syndrome:

Similar to myasthenia gravis, antibodies to calcium channels result in proximal limb weakness, however there is *increased* strength on repeated movement (c.f. myasthenia). *Associated with small cell carcinoma.*

Hypertrophic Pulmonary OsteoArthropathy (HPOA):

Proliferative periostitis (seen on Xray as onion skin appearance of distal long bones) results in joint pain and stiffness at wrists and ankles. *Associated with squamous and adenocarcinomas.*

Investigations

Aims:

The aim of investigation (which is applicable for any carcinoma) is to confirm the diagnosis, to gauge severity by staging and grading the tumour and to identify complications. This gives an idea of prognosis and guides treatment.

In a nutshell, *CXR* is the first investigation, which is followed by a *staging CT*.

Staging = "extent of spread"

T (0-4)	Size and local invasion of tumour
N (0-3)	Extent of lymph node involvement
M (0-1)	Presence of distant metastases

Grading = "degree of malignancy"

Determined by degree of differentiation as seen under the microscope.

FNA is used to biopsy peripheral tumours while *bronchoscopy* is used to biopsy central tumours. If biopsy is not possible, *sputum cytology* may be helpful. Biopsy confirms the histological type of tumour and grade.

Simple	ECG	Arrhythmias
	Early morning urine	SIADH
Lab		
Haem	**FBC**	Anaemia, infection
	INR	Liver mets: ↓ clotting
Chem	U&E, **Calcium**	Hypercalcaemia
	Cortisol, ADH	Endocrine complications
	LFTs	Liver mets: transaminitis
Cytology	Sputum and pleural fluid	
Imaging	**CXR**	90% sensitive Distinct mass (>1-2cm are visible) Lymphadenopathy Complications: Pleural effusion/consolidation/collapse
	CT	Visualisation of tumour **Staging** & operability
	Bone scan	Staging
Invasive	**FNA**	**Bx: Grading**
	Bronchoscopy	**Bx: Grading**

Treatment

Treatment is guided by the type of lung cancer, staging, grading and co-morbidities. Each patient is discussed in a multi-disciplinary team meeting where holistic treatment options are discussed. Important resources include specialist nurses and organisations such as Macmillan cancer relief.

Curative

Surgery ± radiotherapy or chemotherapy

Radical radiotherapy

Only 20% are suitable for surgery with a 5 year survival of 20%.

Surgery is not an option for small cell carcinoma, which has a very poor prognosis. Chemotherapy may prolong survival from an average of 2 to 10 months.

Palliative

Early referral to specialist services such as specialist nurses, pain team, oncology physiotherapists and occupational therapists.

Radiotherapy can be useful in the palliation of breathlessness, cough, haemoptysis, superior vena cava obstruction and chest pain.

Airway/ oesophageal obstructions	Stenting
Pleural effusions	Pleural tap/drain, pleurodesis
Hoarseness	ENT referral
Symptoms from brain metastases	Steroids, radiotherapy
Symptoms from bone metastases	Radiotherapy
Difficulty swallowing	SALT
Weight loss, loss of appetite	Dietician input
Depression	Counselling

Prognosis

Prognosis is poor. Non-small cell carcinoma (without metastases) has a 50% 2 year survival while small cell carcinoma has a 3 month median survival.

GASTROENTEROLOGY

GASTROENTEROLOGY

HISTORY

Cause

Gastrointestinal risk factors

1. Genetics Family history

2. Exposure e.g. *liver disease*: alcohol/ IVDU/ blood transfusion/ sex
e.g. *gastroenteritis*: recent travel, food eaten
Many GI disorders have no identifiable risk factors

Effect

Symptoms and ***severity***

1. Abdominal pain

Site (by quadrants)
Onset
Character
Radiation
Associated symptoms
Timing
Exacerbating/alleviating factors
Severity (from 1 to 10)

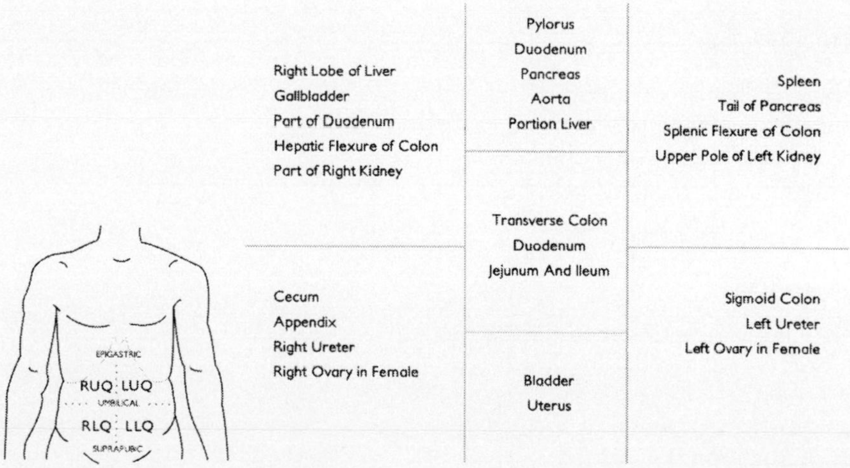

EPIGASTRIC
RUQ LUQ
UMBILICAL
RLQ LLQ
SUPRAPUBIC

Right Lobe of Liver
Gallbladder
Part of Duodenum
Hepatic Flexure of Colon
Part of Right Kidney

Pylorus
Duodenum
Pancreas
Aorta
Portion Liver

Spleen
Tail of Pancreas
Splenic Flexure of Colon
Upper Pole of Left Kidney

Transverse Colon
Duodenum
Jejunum And Ileum

Cecum
Appendix
Right Ureter
Right Ovary in Female

Bladder
Uterus

Sigmoid Colon
Left Ureter
Left Ovary in Female

2. Stool

Constipation/diarrhoea? *This is subjective – "deviation from normal"*

Frequency	What is normal for them?
Urgency	Do they get to the toilet on time?
Consistency	Liquid/ soft/ hard

Colour	Mucus/ blood/ malaena/ steatorrhoea
Smell	Offensive?
Tenesmus	Sensation of needing to defaecate without having anything to pass

3. Rectal bleeding

Quantify	Number of episodes + tinge/ teaspoon/ cupful
Colour	Bright red/ dark brown
Location	Mixed in/ on surface
Other symptoms	Pain, pruritis (↑ likelihood haemorrhoids)
Anaemia	Fatigue/ SOB

4. Nausea and vomiting

Quantify	Number of episodes + amount
Colour	Bilious/coffee-ground/ haematemesis (quantify)

5. Dysphagia

To liquids/solids?
Food sticking high up/low down in throat
Pain on swallowing (odynophagia)
Regurgitation of food

6. Jaundice

Yellow tinge to skin/sclera
Dark urine, pale stool?

7. Constitutional symptoms

Weight loss	How much over how long?
Loss of appetite	Eating and drinking?

Examination

Introduce

Name/age/occupation

Observe

1. **Surroundings**
2. **BMI** Cachectic/ obese?

Hands 1. **Nails** Clubbing*
Leukonychia† (↓albumin), kolionychia (↓Fe)

2. **Palms** Palmar erythema*, Dupuytren's contracture*

Arms Asterixis†/ liver flap =encephalopathy

Face 1. **Eyes**: Anaemia (you only need to look in <u>one</u> eye)
Jaundice†

2. **Mouth**: Aphthous ulcers (Crohn's/ Coeliac disease)
Hepatic foetor†

Neck Lymphadenopathy (notably Virchow's node: Left supraclavicular node indicative of GI cancer)

Chest Gynacomastia* You have to *feel* the breast tissue

Spider naevi* Press to confirm that they blanche & fill from the centre (>5 is significant)

---

*Signs of chronic liver disease

†Signs of decompensated liver disease

Note: Signs aren't necessarily specific e.g. Dupuytren's has various associations including diabetes and phenytoin.

Abdo 1. **Observe** carefully for scars, distension, caput medusa*

Asking the patient to lift head off pillow reveals most hernias

Dilated abdominal veins:

Caput medusa	Flow away from umbilicus (**down** if below umbilicus)
IVC obstruction	Flow towards the heart (all flow **up**)

Caput Medusa vs. IVC Obstruction

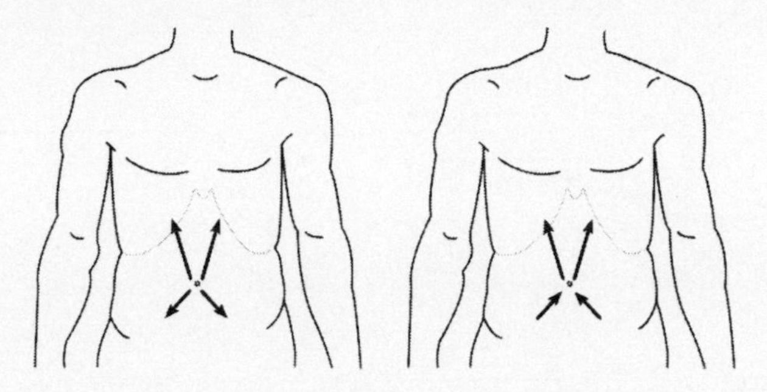

2. **Palpate**

Ask patient if in any pain *before* you touch them. Bend down almost to your knee so that your wrist is straight and level with the patient. Look at their face during palpation for signs of tenderness e.g. wincing.

Superficial =Screen for pain/ rigidity
If present: rebound tenderness/ guarding?

Abdominal Pain

Gallbladder Disease Duodenal Ulcer Acute Pancreatitis Hepatitis Pneumonia Subphrenic Abscess	Peptic Ulcer Oesophagitis Pancreatitis Aortic Aneurysm	Gastric Ulcer Pancreatitis Splenic Rupture Pneumonia Subphrenic Abscess
	Early Appendicitis Small Intestinal Obstruction Acute Gastritis Acute Pancreatitis Ruptured Abdominal Aneurysm	
Appendicitis Inflamed Meckel's Diverticulum Crohn's Disease Ectopic pregnancy Pelvic Inflammatory Disease Salpingitis Ureteric Colic	Acute Urinary Retention Urinary Tract Infection Cystitis Pelvic Inflammatory Disease Ectopic Pregnancy Diverticulitis	Diverticulitis Constipation Ulcerative Colitis Ectopic Pregnancy Pelvic Inflammatory Disease Salpingitis Ureteric Colic

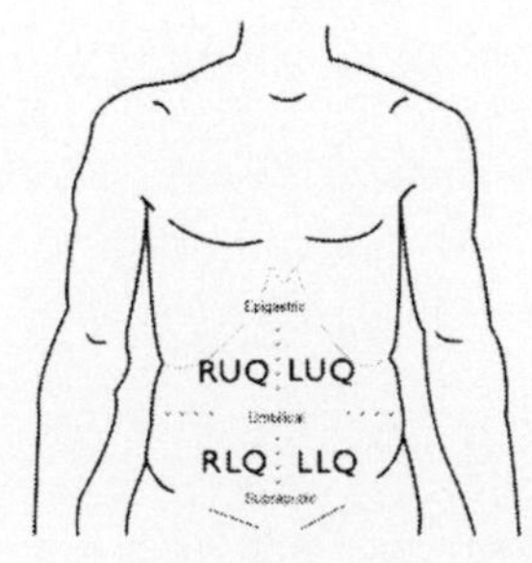

Deep =Feel for masses

Site/ size/ texture/ tender?/ percuss/auscultate

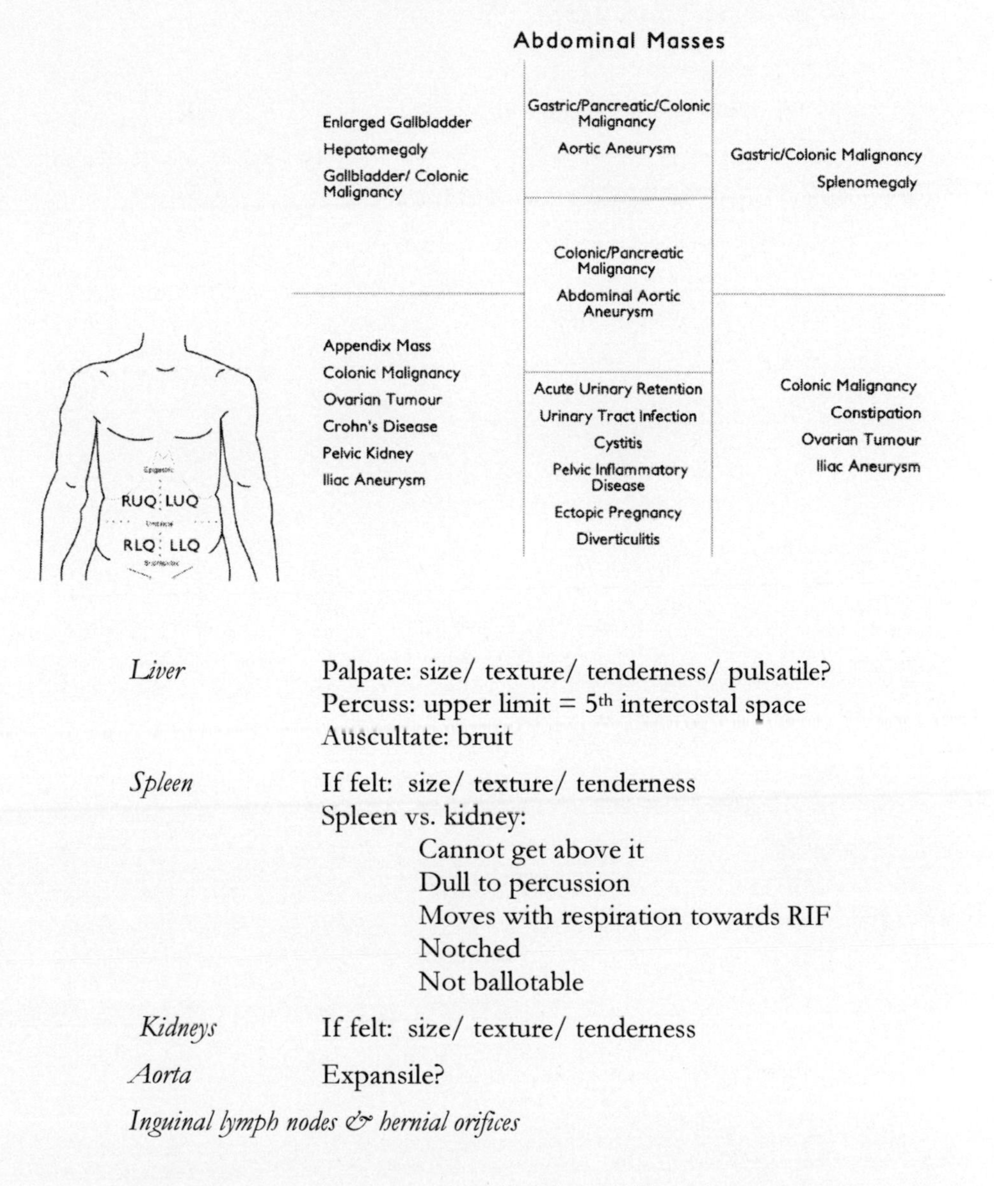

Liver Palpate: size/ texture/ tenderness/ pulsatile?
Percuss: upper limit = 5th intercostal space
Auscultate: bruit

Spleen If felt: size/ texture/ tenderness
Spleen vs. kidney:
- Cannot get above it
- Dull to percussion
- Moves with respiration towards RIF
- Notched
- Not ballotable

Kidneys If felt: size/ texture/ tenderness

Aorta Expansile?

Inguinal lymph nodes & hernial orifices

3. **Percuss**

Pain on percussion = peritoneal inflammation = peritonism
Shifting dullness = ascites† (do not perform this if no dullness!)

4. **Auscultate** over aorta (*above* umbilicus)

Bruit?
Bowel sounds? absent: ileus/ tinkling: obstruction

Legs Pitting oedema†, if present, look for level (ankle/knees/thighs)

"To complete my examination, I would like…"

Examine genitalia/ perform rectal exam/ temperature

THE ACUTE ABDOMEN *(highly unlikely to be seen in practical exams)*

E.g. **Classic appendicitis**

Initially starts off as visceral pain. There is no direct innervation to the inflamed appendix, which results in a vague, **visceral** pain referred to the umbilical area. When the parietal peritoneum becomes irritated, **somatic** innervation of this peritoneum allows the pain to be localised to the right iliac fossa and the nature of the pain becomes sharper. As inflammation develops and the appendix bursts, a peritonitic picture evolves, with worsening in intensity of the pain and **rebound tenderness** (pain on releasing a hand pressed to the abdomen), **guarding** (the *involuntary* contraction of localised abdominal wall muscles to guard inflamed organs from pain upon palpation) and **rigidity** (the *involuntary* contraction of abdominal wall muscles, "generalised guarding"). Generalised peritonitis is a serious condition as it leads to toxaemia and septicaemia.

Visceral pain

Indication of *visceral organ* inflammation
Vague, referred pain
Duller, hard to pinpoint

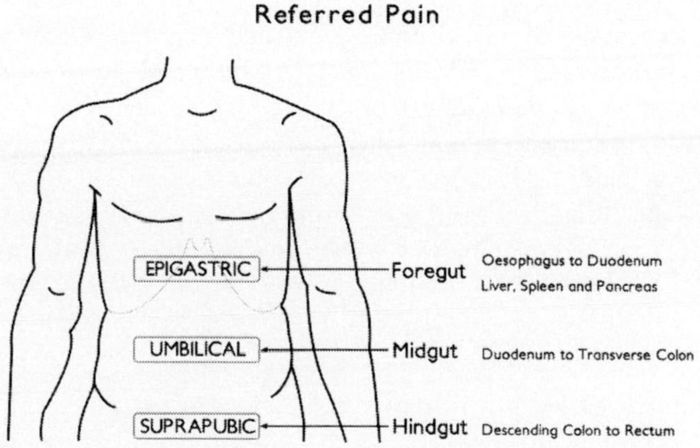

Referred Pain

EPIGASTRIC ← Foregut — Oesophagus to Duodenum, Liver, Spleen and Pancreas
UMBILICAL ← Midgut — Duodenum to Transverse Colon
SUPRAPUBIC ← Hindgut — Descending Colon to Rectum

Somatic pain

Indication of *parietal peritoneal* irritation/inflammation

Sharper, patient able to localise pain, pointing with a finger

Rebound/ percussion tenderness = Localised area of peritonitis
Guarding = Localised area of peritonitis
Rigidity = Generalised peritonitis

Presentation

1. Name, age, occupation

2. Appearance Surroundings/ BMI

e.g. "The patient was alert and comfortable at rest."

3. Peripheral signs: Hands, arms, face, neck and legs

Positives & relevant negatives:

e.g. "His fingernails were clubbed and he had palmar erythema. He had no liver flap, hepatic foetor or signs of jaundice or anaemia; however I found 7 spider naevi on his chest and arms. There was no cervical lymphadenopathy or leg oedema."

Alternatively 'There were no peripheral stigmata of liver disease' is a useful phrase to cover a lot of relevant negatives.

4. Abdomen: Observation, palpation, percussion, auscultation

e.g. "On examination of his abdomen, he had a gridiron scar indicative of previous appendicectomy. His abdomen was soft and non-tender. His liver was smoothly enlarged, extending from the 4th intercostal space to four finger-breadths below the costal margin and non-pulsatile. Neither his spleen nor kidneys were palpable. His aorta was palpable, but not enlarged or expansile and there was no bruit. Other than over the enlarged liver, the percussion note was resonant. On auscultation, bowel sounds were present and normal."

6. Summary

Put everything together with positives & relevant negatives

e.g. "In conclusion, Mr Smith is a 55 year old gentleman with compensated chronic liver disease as evidenced by smooth hepatomegaly and stigmata of liver disease. He has no signs of an acute abdomen. My differential diagnosis includes viral hepatitis.

Note: The differential diagnosis here is large and should be tailored to the most likely depending on patient's background etc.

INVESTIGATION

ABDOMINAL X-RAYS

Describe

Identify the investigation and the patient.

"This is an abdominal radiograph of whoever taken whenever"

From here it's usually best to go on to the most obvious abnormality.

Gas pattern

Look at the bowel gas pattern. Follow the rectum up around the colon, then the small bowel looking for dilated loops of bowel = obstruction.

Small bowel

Centrally placed loops
Valvulae conniventes (complete, extend across bowel wall)

Large bowel

Peripheral loops
Haustrae (incomplete, do not cross whole diameter of wall)

Free air

Best seen in an erect chest X-ray as air under the diaphragm, it may also be seen as Rigler's sign with gas (black) on both sides of the bowel wall = perforation.

Calcification

Vascular	Aneurysms
Renal	Stones (more likely to have calcium containing stones)
Gallstones	Stones (less likely to have calcium containing stones)
Liver/ spleen	TB
Pancreas	Chronic pancreatitis

Soft-tissues and bones

Look out for the hemiarthroplasty/ total hip replacement

Abdominal X-rays are much more exciting when they contain contrast. If so, make sure you identify:

Contrast

Mention that it's been given and *where* it is

e.g. "I can see contrast from the rectum to the caecum"

e.g. "Double contrast study" if air also given (Ba enemas)

Ask for control film if contrast is visible (particularly for KUB)

Type of investigation

Barium swallow (Oesophagus)

Barium meal (Stomach)

Small bowel follow-through (Small bowel)

Barium enema (Colon)

Kidneys, ureters, bladder (KUB)

Endoscopic retrograde cholangiopancreatogram (ERCP)

Percutaneous transhepatic cholangiogram (PTC)

Position of patient Left/right lateral (for barium enemas)

PATHOLOGY

GASTRO-OESOPHAGEAL REFLUX DISEASE (GORD)

Definition

Recurrent acid reflux into the oesophagus due to failure of the lower oesophageal sphincter (LOS) causing persistent symptoms and /or oesophagitis.

Oesophagitis =Inflammation of the oesophagus

Gastro-oesophageal reflux =Reflux of gastric contents, may be normal

Description

Incidence Quite common, prevalence of 5%

Age Any, ↑ w/age, especially over 40 yrs

Sex M:F (GORD)

Geography Worldwide

Aetiology

Low LOS tone

Hiatus hernia

Systemic sclerosis

Following dilatation e.g. for achalasia

Drugs: anticholinergics, bisphosphonates, smoking

↑ *Intra-abdominal pressure*

Obesity, pregnancy

Delayed gastric emptying

Large meals/ fatty meals/ chocolate, coffee, alcohol

Cigarette smoking

Pathophysiology

Reflux of acid into the oesophagus may be due to a hiatus hernia.

Hiatus hernia = protrusion of gastro-oesophageal junction (GOJ) ± stomach above the diaphragm

Remember the LOS has no muscle and depends partly on the diaphragm for tone.

Sliding (80%)

Gastro-oesophageal junction above diaphragm

Rolling

Cardia of stomach above diaphragm. This type should be repaired as there is a risk of strangulation

Mixed

Gastro-oesophageal junction *and* cardia above diaphragm

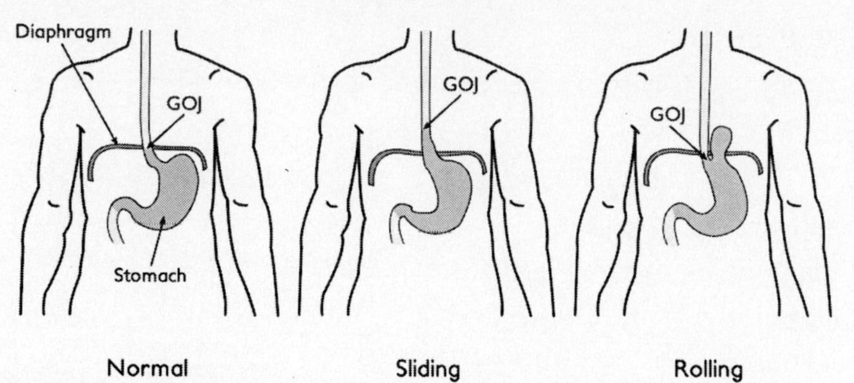

Barrett's oesophagus = a complication of longstanding GORD; squamous to columnar metaplasia occurs in the oesophagus which is pre-malignant for adenocarcinoma. It is diagnosed on biopsy obtained via OGD and may be treated medically with PPIs, laser ablation at endoscopy or surgery.

History and Examination

Symptoms correlate poorly with degree of oesophagitis.

"Positional retrosternal chest pain"

Burning pain "heartburn"

Exacerbated by	Bending/stooping/lying down
	Hot liquids/ alcohol/ spicy food
Alleviated by	Antacids/ milk

When presented with this relatively common disease, you must ask questions and carefully examine to rule out the following:

MI/ angina	Central, crushing chest pain, ±N&V
	Worse on exertion
	Not relieved by antacids, relieved by GTN
Gastric/ duodenal ulcer	Any bleeding/symptoms of anaemia?
GI malignancy	Constitutional symptoms/ odynophagia Metastases?

Investigations

While usually a clinical diagnosis, investigations help rule out other pathology.

Simple	ECG	r/o MI in the acute presentation
Bloods	FBC	Anaemia from bleeding ulcer
Imaging	Ba swallow	Anatomical lesions/motility disorders
		Not sensitive for oesophagitis
Special	**24hr pH monitor** (diagnostic in unclear cases)	
		pH < 4 for $>5\%$ of time monitored
Invasive	**OGD**	Assess severity of oesophagitis
		Biopsy ?Barret's oesophagitis
		r/o peptic ulcer

*Indications for OGD include: (*NICE Guidelines 2005)

Persistent symptoms / symptoms refractory to treatment

Suspicion of cancer (constitutional symptoms/ dysphagia)

Treatment

Conservative

Weight loss

Avoid tight clothes

Raise head of bed at night

Avoid alcohol, heavy meals and precipitating foodstuffs

Eat at least 4 hours before bed

Smoking cessation

Medical

Antacids

Magnesium salts (cause diarrhoea)
Aluminium salts (cause constipation)
Alginates=form "raft" over gastric contents e.g. Gaviscon

H_2 antagonists

Ranitidine

Proton pump inhibitors

Omeprazole, Lansoprazole

Prokinetic agents

Metoclopramide	↑ Gastric emptying
Cisapride	↑ Oesophageal motility

Interventional

Fundoplication

Commonly a laparoscopic procedure, the fundus of the stomach is wrapped around terminal oesophagus following reduction of hernia if present. Complications include inability to belch which may cause bloating

Prognosis

Most GORD is controlled with conservative and medical treatment

Complications of GORD include peptic stricture, ulcers and Barrett's oesophagus leading to adenocarcinoma in 0.5-1% as detailed above.

ACHALASIA

Definition

Achalasia is aperistalsis of the oesophagus and failure of relaxation of the lower oesophageal sphincter.

Description

Incidence Rare

Age Any, rare in childhood

Aetiology Idiopathic/ secondary to malignancy, diabetes

Pathophysiology

Degeneration of mesenteric plexus/ vagus nerve

Dilatation and muscular hypertrophy of oesophagus

History and Examination

Intermittent dysphagia *of both liquids and solids*

Regurgitation of food

Retrosternal chest pain (GORD)

r/o: oesophageal cancer/ benign stricture/ oesophagitis

Investigations

Manometry	Aperistalsis of LOS
Barium swallow	Dilatation of oesophagus and beak constriction
OGD	Biopsy to exclude carcinoma

Treatment

Interventional

Endoscopic

Balloon dilatation

Botulinum toxin injections

Complications include oesophageal perforation and GORD

Surgical

Longitudinal myotomy of gastro-oesophageal junction

Prognosis

These patients have a slightly increased risk of oesophageal cancer.

Oesophageal Cancer

Definition

Oesophageal cancer can be squamous or adenocarcinoma. These differ in aetiology and affect different parts of the oesophagus.

Description

Incidence	8th mc cancer Most rapidly increasing cancer in western world (Adenocarcinoma)
Age	60-70yr olds
Sex	M>F
Geography	China, Iran

Aetiology/ Risk factors

Squamous cell	Smoking Alcohol Achalasia
Adenocarcinoma	Barrett's oesophagus = 30x risk of CA GORD

Pathophysiology

Squamous cell	Upper 2/3 oesophagus, mc **mid oesophagus**
Adenocarcinoma	Lower 1/3 oesophagus (As it's due to acid reflux!)

Both result in narrowing and ulceration of the oesophagus, and can spread by:

Direct invasion

Lymph nodes

Bloodstream → lung, liver, brain

Oesophageal cancer

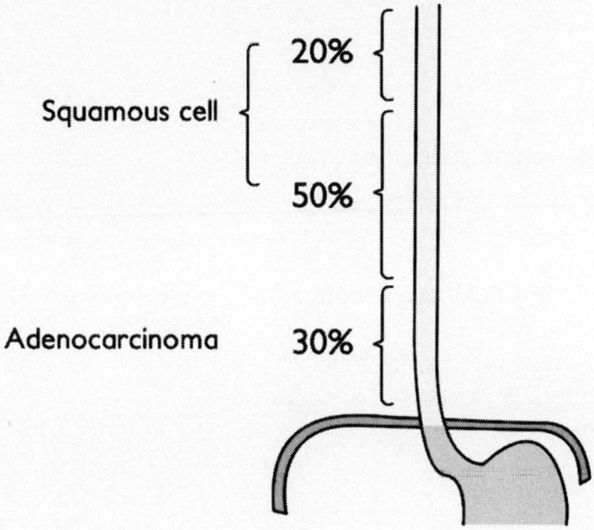

Squamous cell 20%
50%
Adenocarcinoma 30%

History and Examination

CAUSE	Hx of risk factors
EFFECT	
Tumour	**Progressive dysphagia** solids → liquids Cough and hoarseness (upper 1/3) Aspiration pneumonia
Constitutional	Weight loss, loss of appetite
Local infiltration	Retrosternal chest pain Lymphadenopathy
Metastatic	Lung, liver, brain (rarer)

Investigations

Imaging	Barium swallow	90% sensitivity Best for "high dysphagia" Still require OGD for Bx
	CT	Tumour volume Local invasion Staging (TNM)
Invasive	**OGD**	Visualisation + biopsy

Treatment

Curative

Oesophagectomy

Wide resection margins
Radical lymph node excision

Endoscopic mucosal resection (specialist centres only)

Palliative (mc)

Multi-disciplinary team approach
Nutritional support
Dilatation/ stenting of stricture
Photocoagulation/ local alcohol injection
Chemotherapy ± radiotherapy

Prognosis

Poor: 10% 5 year survival rate (60% if confined to mucosa)

PEPTIC ULCER DISEASE

Definition

Ulceration of stomach or proximal duodenum usually associated with H. pylori or NSAID use.

Understand that gastric ulcers are quite different to duodenal ulcers but difficult to distinguish on the basis of symptoms.

Description

Incidence	Very common	
	Gastric Ulcer (GU)	Lifetime incidence 5%
	Duodenal Ulcer (DU)	Lifetime incidence 15%
Age	↑ with age	
Sex	**GU** M=F	
	DU M>F 4:1	
Geography	Northern UK > Southern UK	

Aetiology and Pathophysiology

↑ Acid secretion	**H. pylori** (90%) 90% of GUs 80% of DUs
	Zollinger-Ellison syndrome (pancreatic gastrinoma)
↓ Mucosal protection	**NSAIDs** (10%)

Risk Factors

Smoking	GU/DU	↓ Healing
Blood group O	DU	
Stress	GU	Curling's ulcer (burns) Cushing's ulcers (neurosurgery) These are now rare due to PPI use

History and Examination

The history is most useful. Epigastric tenderness may be present on clinical examination.

CAUSE

H. pylori	Extremely high prevalence of 50-60% No clinical features
NSAIDs	Patients with chronic pain: e.g. rheumatoid arthritis, headache, lumbar pain

EFFECT

Epigastric pain	**Visceral pain** On eating = classic GU When hungry = classic DU Relieved by antacids
N&V	Vomiting is unusual but may relieve symptoms

Complications:

Haemorrhage	Haematemesis/ coffee ground vomit Melaena = blood proximal to caecum Anaemia

Rx: ABC → Gastroscopy: visualise bleed → adrenaline/ diathermy

Perforation	***Surgical emergency*** Acute abdomen (peritonitis) ± shock Air under diaphragm on erect CXR Rigler's sign on AXR

Rx: ABC → conservative / laparotomy: undersew ± omental patch

Investigations

An important aim of investigation is to rule out cancer.

Simple	ECG	r/o MI
Bloods	**FBC**	Anaemia due to haemorrhage
	WBC	Lymphoma
	Amylase	Pancreatitis
	LFTs	Cholecystitis
	Serum gastrin	Pancreatic gastrinoma
Imaging	**Erect CXR**	If suspicion of perforation
Special	**Urea breath test**	Detects H pylori Very sensitive and specific Patient must be off PPI
	OGD	**Visualise and biopsy ulcer** Rapid urease test for H pylori Histology to r/o cancer

All gastric ulcers must be biopsied (duodenal malignancies are rare)

Treatment

The aim of treatment is to reduce acidity of gastric contents and treat complications.

Conservative

Avoid NSAIDs

Smoking cessation

Medical

Proton pump inhibitors (PPIs)	Lansoprazole/ omeprazole
H_2 antagonists	Ranitidine/ cimetidine
Prostaglandin inhibitors	Misoprostol

H. pylori eradication therapy:

"Triple therapy" = PPI + 2ABx

1. PPI
2. Clarithromycin
3. Metronidazole/amoxicillin

Following eradication therapy, gastric ulcers must be endoscopically visualised at six weeks to rule out cancer.

Interventional

Due to the effectiveness of medical treatment, surgery is now only used to treat complications.

Now obsolete surgery to reduce acidity includes:

Partial gastrectomy
Total/selective vagotomy

Complications:

Recurrent ulcer
Dumping syndrome, due to rapid gastric emptying:
Sweating, nausea, palpitations, hypoglycaemia after eating

Prognosis

Prognosis is good. H pylori has been designated as *carcinogenic* but eradication, which is 90% effective, affords 100% cure. Duodenal ulcers **do not** lead to carcinoma.

GASTRIC TUMOURS

Definition

Neoplasms of the stomach are usually malignant adenocarcinomas.

Description

Incidence	Relatively common, 6th most fatal cancer in UK
Age	↑ with age, peak 50-70 years
Sex	M>F 3:1
Geography	Common in Japan

Aetiology & Pathophysiology

There is a strong link between H. pylori and gastric adenocarcinoma. Gastritis leads to dysplasia, a precursor for carcinoma. *H. pylori is carcinogenic.*

Adenocarcinomas are most commonly found in the antrum, look irregular, ulcerate and have rolled edges. Generalised submucosal spread is termed *linitis plastica* as it results in a rigid stomach, best seen on barium meal.

Risk Factors

Congenital	**Family history**
	Blood group A
	Pernicious anaemia (due to atrophic gastritis)
Acquired	**H. pylori**
	High salt diet
	Smoking

History and Examination

Patients usually present with advanced disease (therefore poor prognosis)

CAUSE	Hx of risk factors
EFFECT	
Tumour	**Epigastric pain (mc)** -relieved by food and antacids **Epigastric mass + tenderness (50%)** Anaemia Vomiting, especially if pyloric tumour Dysphagia if fundus involved
Constitutional	Weight loss, anorexia
Local infiltration	Virchow's node / Troisier's sign
Metastatic	Liver, bone, brain, lung (30%)
Paraneoplastic	Dermatomyositis Acanthosis nigricans

Investigations

Blood	FBC	Anaemia
	LFTs	Liver mets
Imaging	Ba meal	90% sensitivity
	CT	Staging, resectability T= extent of tumour invasion N= lymph node involvement M= metastatic disease
Invasive	**OGD**	Visualisation + biopsy

Although Ba meal and increasingly OGD screening exists in Japan where the incidence of early gastric cancer is high, there are no screening programmes in the UK at present. Gastric cancer screening is a good example of lead time bias (where early detection merely increases the duration of the patients' awareness of their disease without reducing their mortality or morbidity).

Treatment

Curative

Surgery is the only form of curative treatment.

Palliative

Multi-disciplinary team approach: Specialist nurses, dietician etc.

Surgery

Chemotherapy (adjuvant)

Prognosis

The prognosis is poor in the UK with 10% overall 5 year survival. This increases to 50% if the tumour is operable.

ACUTE PANCREATITIS

Definition

Acute inflammation of the pancreas which can be aseptic or septic with variable involvement of local tissues and remote systems.

Description

Incidence	Relatively uncommon, but one to watch out for as 25% require ITU
Age	Any, peaks at 40-60yrs
Sex	M:F overall, but depends on aetiology
	Young males → due to alcohol
	Older females → due to gallstones

Aetiology

The most common causes are:

Gallstones (45%)

Alcohol (25%)

Idiopathic (20%)

"GAIT SMASHED" is a variation of "get smashed" so that first three causes are in order of frequency:

Gallstones

Alcohol

Idiopathic

Trauma

Steroids

Mumps (and other infections)

Autoimmune

Scorpion bite (make sure this is the LAST cause you offer, if at all!)

Hyperlipidaemia/ hypercalcaemia, hypothermia, hypotension

ERCP

Drugs: Thiazide diuretics, azothioprine, sulfonomides, anti-retrovirals

Pathophysiology

Inflammation of the pancreatic acini results in release of enzymes that have destructive effects. Oedema, haemorrhage and necrosis then follow to varying degree. Mild pancreatitis involves self limiting minimal organ dysfunction with an uneventful recovery. Severe pancreatitis involves pancreatic necrosis, local complications and organ failure.

In pancreatic necrosis there are diffuse or focal areas of non-viable pancreas with fat necrosis. Liberated pancreatic enzymes start to digest the pancreas and other retroperitoneal tissues. Lipases cause fat necrosis, proteases cause tissue breakdown and other enzymes, including elastase, damage blood vessels, resulting in haemorrhage.

There can be massive third space loss when fluid accumulates in the gut, peritoneum and retroperitoneum. In severe cases there is a systemic inflammatory reaction with the activation of complement, kinin and coagulation/fibrinolytic cascades. ARDS and multiple organ dysfunction (MOD) may ensue.

History and Examination

CAUSE: GAIT SMASHED

EFFECT:

Pain Epigastric
Radiates to back, relieved sitting forward
Tender, distended abdomen

Vomiting Intermittent, bilious

Ecchymosis Rare, indicates severe haemorrhagic pancreatitis

Grey Turner's sign: Bruising around flanks
Retroperitoneal bleeding tracks superficially

Cullen's sign: Bruising around umbilicus
Blood tracking down ligamentum teres

Jaundice

Shock to varying degrees:

Remember percentages of blood loss are like tennis scores, running from Love (0%) to Forty (40%+) = Game! (See table below)

The first real clinical signs are at stage II shock with an increase in heart rate and diastolic BP (which is the same as ↓ pulse pressure). These maintain BP, which falls late in the game at stage III shock.

Stage	Blood loss	HR	BP	Urine output	Mental state
I	**0-15%** 750mls	Normal	Normal	Normal	Mildly anxious
II	**15-30%** 750ml-1.5L	**Increased** >100bpm	**Increased diastolic**	Decreased 20-30ml/hr	Anxious
III	**30-40%** 1.5-2 L	Increased >120bpm	**Decreased**	Decreased <20 ml/hr	Confused
IV	**40%+** **(Game!)** >2L	Increased >140bpm	Decreased	None	Lethargic LOC

Investigations

The aim of investigation is to make the diagnosis, gauge severity and obtain prognostic factors.

Simple	ECG	
	ABGs	Severity: Metabolic acidosis/ respiratory failure
Blood	**FBC**	Anaemia, inflammation
	U&E	Dehydration, severity
	LFTs	Severity, jaundice
	Amylase	Diagnosis, *not* severity
	CRP	Severity, useful for monitoring progress
	Ca, LDH	Severity
Imaging	**CXR**	Excludes perforation (air under the diaphragm)
		Respiratory complications (ARDS, pleural effusion)
	AXR	Exclude perforation
		Calcified pancreas: Acute on chronic pancreatitis
		Sentinel loop = duodenal ileus
	US	Identify gallstones, visualise pancreas (may be difficult)

CT	Diagnosis, baseline imaging Screen for complications: Haemorrhage Necrosis Pseudocyst

Prognostic indicators

Amylase is **not** a prognostic indicator. It is useful for diagnosis as it is raised in 80% of patients 2-12 hours after onset of symptoms. An amylase greater than 1000 confirms the diagnosis. It is **not** specific to pancreatitis and may be raised in other "ruptures" such as perforated peptic ulcer, ruptured AAA and ectopic pregnancy as well as DKA and MI.

Several scores and criteria exist for severity and mortality. The Glasgow Criteria is used in the UK and it has been validated for both alcoholic and gallstone pancreatitis. You may also have heard of the Ranson criteria which are based on scores on admission and at 48 hours.

The Glasgow Criteria "PANCREAS"

PO_2	<8
Age	>55
Neutrophils	>15
Calcium	<2
U**R**ea	>16
Enzymes: LDH	>600
Albumin	<32
Sugar: glucose	>10

3+ points within the first 48 hours = severe acute pancreatitis.
CRP of >150 @ 48hrs is also a good indicator of a severe attack.

Treatment

The aims of treatment are to resuscitate, support and monitor the patient, treat the cause and any complications.

Resuscitate

ABCDE

Support

Monitor, monitor, monitor! These patients can go off very quickly.

Admit to HDU/ITU

Analgesia

Careful fluid management (remember extensive 3rd space loss)

Oxygen (PO_2 is a prognostic indicator)

NBM to rest pancreas + TPN/ Nasojejunal feeding (debatable!)

NG tube if vomiting

Insulin sliding scale (remember the pancreas isn't working anymore)

Correct metabolic disturbances: Ca, alcohol withdrawal

Antibiotics (not always necessary)

Interventional

Removal of stones + sphincterotomy may ↓ complications

Necrosectomy (debridement of necrotic pancreas)

Peritoneal lavage

Laparostomy (abdomen left open in ITU setting)

Drainage of abscesses/ pseudocyst

Prognosis

Mortality of 10%, though depends on severity. Lots of potential for complications:

Local

Necrosis: 50% of these progress to infection (this trebles the risk of death)

Abscess formation

Haemorrhage

Fistula

Colonic infarction

Pseudocyst: 2-4 weeks post attack

A pancreatic pseudocyst is a collection of pancreatic fluid in lesser sac surrounded by granulation tissue. Features include persistently raised amylase, epigastric pain ± mass.

Ix: Abdominal US/ CT
<6cm resolve spontaneously
>6cm require treatment: ↑risk of rupture or obstruction

Rx: US/ CT guided drainage
Cystogastrostomy (cyst drained into posterior stomach)

Systemic

Respiratory: ARDS, pleural effusion, atelectasis

Myocardial depression

Renal failure

DIC

Metabolic: Hypocalcaemia, hypomagnesaemia
Hyperglycaemia
Metabolic acidosis

CHRONIC PANCREATITIS

Definition

Prolonged inflammation and fibrosis of the pancreas resulting in chronic pain and pancreatic endocrine and exocrine dysfunction.

Description

Incidence Not uncommon, incidence increasing

Age Mean age at diagnosis is 45 years

Sex M>F 2:1 overall, but depends on aetiology
Males → Alcohol
Females → Idiopathic

Aetiology

The most common causes are:

Alcohol (75%)
Idiopathic
Obstructive (congenital)
Cystic fibrosis

Pathophysiology

Chronic inflammation results in fibrosis, calcification and cyst formation. Dilatation of pancreatic ducts may cause pain and there is pancreatic dysfunction. The following functions of the pancreas are lost:

Exocrine: *Digestive enzymes*
Trypsinogen
Chymotrypsinogen

Endocrine: *Hormones*
Insulin
Glucagon
Somatostatin
Pancreatic polypeptide
Serotonin

In addition, these patients often have concurrent social problems due to alcoholism and reliance on strong analgesics.

History and Examination

CAUSE: As above

EFFECT:

Pain Chronic, epigastric, radiates to back
Often rely on strong opiates

↓ Exocrine function

Weight loss
Decreased appetite
Steatorrhoea
Malabsorption

↓ Endocrine function

Diabetes: polyuria/ polydipsia

Investigations

The aim of investigation is to make the diagnosis and exclude malignancy and gallstones.

Blood	**FBC**	Inflammation/ infection
	LFTS	Associated complications
	CA 19-9	CA pancreas

	CRP	Inflammation/ infection (Amylase/ lipase is unhelpful)
Imaging	**AXR**	Calcified pancreas
	US	Exclude gallstones
	CT	Inflammation, exclude cancer
Invasive	ERCP	Information on ducts, obstruction?

Treatment

Conservative

Stop drinking/ smoking
Address social problems
Pain clinic
Dietary advice: healthy, small, frequent meals

Medical

Pancreatic enzyme supplementation
Treat diabetes

Interventional

Relieve blockage: ERCP + stent/ cholecystectomy
Drain cysts
Pancreatic resection (Whipple's procedure)

Prognosis

20 year mortality of 50%, mostly due to complications of alcohol, diabetes or malabsorption.

PANCREATIC CANCER

Definition

Most typically adenocarcinoma of the head of the pancreas, pancreatic cancer carries a very poor prognosis.

Description

Incidence	Relatively uncommon but increasing incidence
Age	40+ but mostly disease of old age
Sex	M>F 1.5:1
Geography	Western

Aetiology

As for most cancers, causes are a variable combination of genetic predisposition and environmental triggers.

Risk Factors

Smoking, alcohol, high fat diet

Diabetes

Chronic pancreatitis

Pathophysiology

Adenocarcinoma (90%) affecting the head of the pancreas (80%) which invades local structures, eventually obstructing the bile ducts and causing obstructive jaundice. Spread occurs via lymphatics and may metastasise via the blood to liver and lungs.

History and Examination

CAUSE: Family history, risk factors as above

EFFECT:

Tumour	Unfortunately the tumour itself has little effect, and is picked up late once there is local invasion to cause symptoms: malabsorption, diabetes
Constitutional	Weight loss, decreased appetite
Local	Painless, obstructive jaundice + palpable gallbladder **Courvoisier's law:** "jaundice with a palpable gallbladder is unlikely to be due to stones"
	Pain -if invades coeliac plexus
Metastatic	Lymphatic
	Haematogenous → liver, lungs

Investigations

Blood	FBC	Anaemia
	LFTS	Obstructive picture: ↑ bilirubin, ALP>>>AST/ALT
	CA 19-9	90% sensitive (also raised in cholangiocarcinoma)
Imaging	US	Visualise cancer and lymphadenopathy
	CT	Diagnose and stage disease
Invasive	ERCP	Cytology brushings Biopsy of periampullary lesions

Treatment

Curative

Only 20% of pancreatic cancers are resectable on presentation.

Whipple's procedure for head of pancreas tumours

Resection of:

Head of pancreas
Duodenum
Distal stomach
Gall bladder

3 anastamoses to jejunum: pancreas, stomach and bile duct

Distal pancreatectomy for tail of pancreas tumours

± splenectomy

Palliative

Most cases are unsuitable for resection due to their size (>2cm) and local invasion.

Conservative
Multidisciplinary team involvement, including specialist cancer nurses

Medical
Analgesia, anti-emetics

Interventional
Relief of obstruction: stenting via ERCP/ percutaneous drainage
Bypass surgery

Prognosis

Prognosis is abysmal. 5 year survival is less than 5%, most die within 3-6 months of diagnosis.

GALLSTONES & BILIARY TREE

Definition

It is important to understand the terminology as biliary diseases have very different presentations, prognoses and treatments.

Gallstones = stones that form in biliary system
Bile = bile salts + lecithin + cholesterol + pigment + water

To simplify: obstructed tubes become inflammed, therefore think of it as obstruction alone, or with superadded inflammation.

GALLBLADDER

Obstruction

Biliary colic (confusingly, not technically a colic)

= Pain caused by muscle spasm of the gall bladder/ducts against an obstructing stone in neck of gallbladder / cystic duct/common bile duct: severe, *constant*, crescendo pain.
(≠ Colic: visceral pain that *comes and goes* due to peristalsis of obstructed viscus.)

Inflammation

Acute cholecystitis = inflammation (chemical, with or without superadded bacterial infection) of the *gallbladder*. Often preceded by biliary colic, usually due to a gallstone obstructing the emptying of the gallbladder.

Empyema = pus in gallbladder
Mucocoele = *sterile* enlargement of gallbladder which may be massive.

COMMON BILE DUCT

Obstruction

Stone in CBD = billiary colic with *obstructive jaundice*:
↑ bilirubin and ↑ ALP>>↑ AST

Inflammation

Ascending cholangitis = infection of *biliary tree* following CBD gallstone

MISCELLANEOUS

Gallstone ileus = *rare* cause of small bowel obstruction due to large gallstone fistulating through the gallbladder to the duodenum, causing obstruction of the small bowel.

Primary sclerosing cholangitis = chronic, *autoimmune*, choleostatic liver disease due to inflammation of *intra/extra hepatic bile ducts*. Often confuses students, has nothing to do with gallstones. Associated with IBD, M>F. and with anti-smooth muscle antibodies and P-ANCA. Slowly progressive, it leads to liver cirrhosis and decompensation. Liver transplant is the only curative treatment.

Description

Incidence	Very common
	15% of population >65yrs (mostly asymptomatic)
Age	Any, ↑ w/age
Sex	F>M
Geography	Scandinavia and South America have higher prevalence

Aetiology

Similar to Virchow's triad for thrombus:

Composition	Lithogenic bile
Stasis	Fasting, TPN
Wall	Abnormalities of epithelium

Lithogenic bile

↑ Cholesterol	Obesity, hyperlipidaemia, diabetes
↑ Pigment	Haemolytic anaemia, sickle cell disease
↓ Bile	Loss of terminal ileum: ↓ enterohepatic circulation
↑Oestrogen	COC, pregnancy, multiparity

Risk Factors

Classically: "Fat, Female, Forty, Fertile"

Pathophysiology

Composition of stones:

Mixed (80%)	Faceted, multiple, 10% radio-opaque
Cholesterol (15%)	Large, solitary, yellow
Pigment (5%)	Dark, soft, mulberries (associated with haemolitic anaemia)

History and Examination

CAUSE: Risk factors as mentioned above

EFFECT:

Biliary colic:

Hx: Right upper quadrant pain ± radiating to back

Severe, continuous, making patient *writhe/ restless*

Classically occurs after a fatty meal

Associated with flatulent dyspepsia, vomiting and sweating

OE: *No abdominal tenderness*, soft abdomen.

Acute cholecystitis:

Hx: Starts with biliary colic then progresses

RUQ pain radiates to back and between scapula
Patient *stays still* due to localised peritonitis
Low grade pyrexia due to inflammation

OE: *Tender RUQ* / guarding/ Murphy's sign

Murphy's sign = sudden, sharp pain during inspiration as gallbladder palpated in mid-clavicular line, under right costal margin. It is only positive if the same test is negative on the left.

Ascending cholangitis

Hx: Charcot's triad

Pain	=Biliary colic
Jaundice	=Obstructive: dark urine, pale stools
Fever/rigors	=Involuntary shivering due to high fever

OE: Tender RUQ, fever, tachycardia, depending on severity

Investigations

Simple	Urine dipstick	
Blood	**FBC**	**Infection**, anaemia
	CRP	Inflammation
	LFTs	**Obstruction:** ↑bilirubin, ↑ALP
Imaging	**USS**	Visualise stone Inflamed, thickened gallbladder Dilated bile ducts: CBD >6mm
	ERCP/ MRCP/ PTC	Visualise/ assess obstruction

Treatment

Conservative

Fluid resuscitation + *analgesia* + *NBM* (biliary colic requires no more than this)

Antibiotics (cef + met) for cholecystitis/CBD stone to prevent ascending cholangitis. Ascending cholangitis is serious and should be treated aggressively.

Interventional

ERCP for urgent treatment of CBD stone with infection

Acute cholecystectomy required if peritonitic, ↑ fever /tachycardia

80% of cholecystitis can be treated conservatively resolving over 1-2 days with cholecystectomy (open or laparoscopic) performed ~ 2 months later.

Prognosis

Complications of gallstones:

Biliary colic → Acute cholecystitis → empyema/ mucocoele

CBD stone → Ascending cholangitis

Pancreatitis if obstructing the outflow of the pancreatic duct

Gallstone ileus (rare)

Gallbladder CA (rare)

COLORECTAL CARCINOMA

Definition

Malignant neoplasm of the colon/rectum, usually adenocarcinoma.

Description

Incidence	3rd most common cancer in the UK Incidence is increasing
Age	↑ w/age, especially >50 years. Peaks at 70 years
Sex	M:F (Rectal: M>F)
Geography	Western disease

Aetiology

Genetic

Familial adenomatous polyposis coli (FAP)
Mutation in APC gene (autosomal dominant)
Inevitable progression from multiple polyps to cancer
Treated with total colectomy with ileo-anal reconstruction (pouch)

Hereditary non-polyposis colon cancer (HNPCC)
10% of colon malignancies
Mutation in DNA mismatch repair genes

Environmental

Diet: high fat, low fibre
Smoking
Crohn's, ulcerative colitis >10 years
Radiation

Pathophysiology

Macropathology

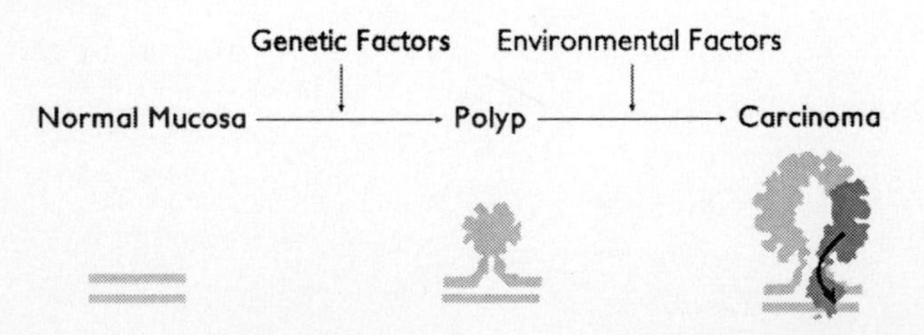

Adenomas (sporadic or familial) are a pre-malignant lesion, which may become cancerous following environmental triggers.

Micropathology

Duke's classification:		**5 year survival**
A	Confined to mucosa	90%
B	Breached muscle layers	60%
C	Lymph node metastasis	30 %
D	Distant metastases	5 %

<u>History and Examination</u>

CAUSE: Family history, risk factors as above

EFFECT:

Tumour	Change in bowel habit Tenesmus (desire to defaecate but no stool) PR bleed/ chronic anaemia Mass
Constitutional	Weight loss, decreased appetite
Local	Obstruction, perforation, ureteric obstruction
Metastatic	Lymphatic Haematogenous → liver (lung, kidney, bones are rarer) Transcoelemic → ovaries

Presentation: Right sided tumours tend to bleed (chronic anaemia) but don't obstruct until they are large as the stool is soft at this point. In contrast, left sided tumours tend to obstruct as by this stage the stool is semi-solid and the lumen is smaller.

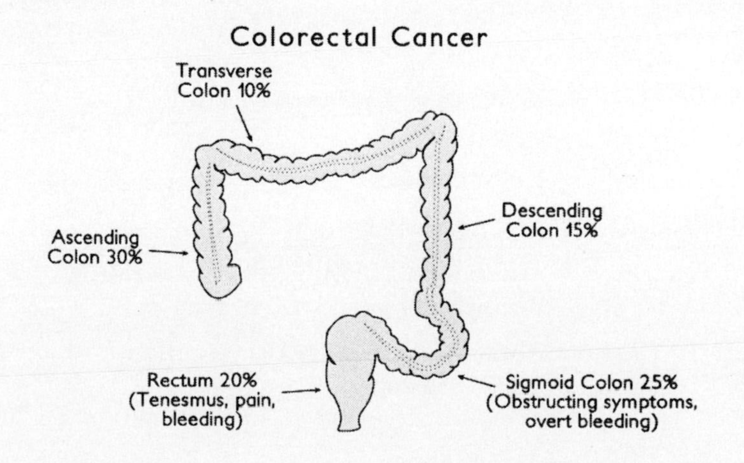

Investigations

Simple	**PR**	
	Faecal occult blood	Screening
Blood	**FBC**	Anaemia
	LFTs	Liver mets
	CEA	Carcinoembryonic antigen Monitor treatment/ recurrence Raised late in disease
Imaging	Liver US	Metastasis
	CXR	Metastasis
	Ba enema	If colonoscopy not possible
	CT	Staging, resectability
Invasive	Sigmoidoscopy	Visualisation + biopsy
	Colonoscopy	Visualisation + biopsy

Barium enema and colonoscopy are the only investigations where the entire colon is investigated. They are important because of the possibility of synchronous neoplasms (more than one tumour present). Colonoscopy is the gold standard as biopsies can be taken for diagnosis and grading.

Colorectal CA screening (colonoscopy, faecal occult blood) may be used for high risk patients such as FAP, strong family history, pervious history of poyps/ colon CA, UC >10 years. National screening was started in 2006 in some parts of the country and is being rolled out. FOB is being offered to all aged 60-69 with colonoscopy offered to positives.

Treatment

Curative

Duke's A, B & C		Resection +/- adjuvant therapy if B/C (radio/chemo)
Right-sided		**Right hemicolectomy**
Left-sided		**Left hemicolectomy**
Rectal	*High*	**Anterior resection** Removal of sigmoid and rectum
	Low	**Abdominoperineal (AP) resection** + end colostomy Removal of sigmoid, rectum and anus

Hemicolectomies and anterior resections may be followed by:

Reanastamosis
Defunctioning stoma with future reanastamosis
Permanent stoma

Palliative

Obstruction/ bleeding	Resection Hartman's procedure: Emergency resection + colostomy
Local disease/recurrence	Radio/chemotherapy
Hepatic metastases	Solitary lesions may be resected Chemotherapy for diffuse metastasis

Prognosis

Duke's A has a good response to treatment; however, few are caught at this stage as patients are asymptomatic. This is a good argument for screening which has not been brought into practice other than for high risk patients as detailed above.

DIVERTICULAR DISEASE

Definition

Diverticula are **outpouchings of viscera**.

Meckel's diverticula are congential and "true" diverticula as all layers of wall are involved. Diverticulosis and pharyngeal pouches are acquired and termed "false" diverticula as only the mucosa and submucosa are involved.

Note: Many diverticu**la**, one diverticu**lum**, while **diverticulitis** is inflammation of a diverticulum

Description

Incidence	Common, 50% of 50 year olds, 60% of 60 year olds etc.
Age	↑ w/age, rare <35 years
Sex	F>M
Geography	Western disease

Aetiology

Risk factors

Increasing age
Female
Obese
Low fibre diet

Saint's triad= diverticula + hiatus hernia + cholelithiasis

Pathophysiology

Increased luminal pressure results in herniation of mucosa where the bowel walls are weaker due to penetration of blood vessels. This can potentially happen anywhere along the GI tract. However, the colon is the commonest site, of which the sigmoid colon is the most affected.

History and Examination

Asymptomatic (90% detected incidentally on Ba enema/ colonoscopy)

Symptoms are minor and include changing bowl habit and intermittent LIF pain.

Complications include:

Diverticulitis
Faeces obstruct neck of diverticulum and result in inflammation ± infection:
"Left-sided appendicitis"
LIF pain + guarding ± mass
Constipation/ diarrhea
Systemic symptoms: fever, tachycardia

Haemorrhage
Faecolith erodes through wall of blood vessel
PR bleeding:
Acute: May need emergency resection (Hartman's)
70% resolve with conservative treatment
Chronic: Fe deficiency anaemia

Perforation
Results in peritonitis: rigidity ± shock

Fistula
An abnormal communication between two hollow organs or a hollow organ with the exterior.
Colovesicular fistula (colon→bladder): pneumaturia/faecouria/haematuria

Strictures
Obstruction may result.

Investigations

Simple	**PR**	CA
	Urine dipstick	UTI/ fistula
	Stool mc&s	
Blood	**FBC**	Anaemia
	ESR	May be raised
	Clotting	
	Cross match	May need transfusion
	Culture	If septic
Imaging	**Ba enema**	Visualise extent r/o CA (not done acutely)
	CT (acutely)	abscess/ perforation/ fistula
Invasive	**Endoscopy**	Visualization r/o CA (not done acutely)

Treatment

Acute

ABC

Supportive : fluids, NBM

Antibiotics if infected (cef+met), analgesia as required

Optimise for surgery, indicated by complications such as perforation, haemorrhage or stricture; repeated attacks or immunosuppression. Hartman's procedure may be necessary, though most are managed conservatively

Chronic

High fibre diet, bulking agents

Elective sigmoid resection and anastamosis

Prognosis

Most diverticula are asymptomatic.

70% of acute episodes will only require supportive treatment. Acute episodes don't always recur therefore diverticula do not necessarily require resection.

Inflammatory Bowel Disease

IBD = Umbrella term for Crohn's disease and ulcerative colitis

Note: IBS≠IBD (Irritable bowel disease vs inflammatory bowel disease)

Crohn's Disease

Definition

Chronic, transmural inflammation affecting any part of the gut from mouth to anus with unaffected bowel between areas of active disease.

Description

Incidence	Rising and variable (UC>CD)
Age	Any, peaks at 20-40
Sex	M:F
Geography	Worldwide. ↑ West, Whites, Jews

Aetiology

Risk Factors

Family history: ↑ Concordance monozygotic twins
Smoking
Autoimmune disease
?Diet, ↑ sugar intake

Pathophysiology

Macropathology

Thickened, narrowed bowel
Cobblestone appearance due to deep ulcers and fissures
Skip lesions (i.e. lesions not continuous)
Terminal ileum most commonly affected

Micropathology

Transmural inflammation
Granulomas (non-caesiating)

History and Examination

Recurrent relapses of:

Diarrhoea, abdominal pain

Acute RIF pain ± mass, may mimic appendicitis

Aphthous ulceration of mouth

Perianal disease (remember "*from mouth to anus*")

Systemic features:

Constitutional	Fever, malaise, weight loss, clubbing
Eyes	Conjunctivitis episcleritis, iritis
Joints	Arthralgia, sacroilitis, ankylosing spondylitis
Skin	Erythema nodosum, pyoderma gangrenosum
Liver	Fatty liver, gallstones, sclerosing cholangitis
Renal	Stones

Note:

Erythema nodosum

An acute, nodular, erythematous eruption, usually limited to the extensor aspects of the lower legs. Presumed to be a hypersensitivity reaction, it may occur in association with several systemic diseases or drugs.

Inflammatory:	Sarcoidosis, IBD
Infective:	Streptococcus, TB, Chlamydia
Malignancy:	Lymphoma, leukaemia
Drugs:	Sulphonamides, COC

Pyoderma gangrenosum

Massive neutrophillic infiltration results in tissue necrosis, initially seen as pustules which may progress to deep ulcers, usually on the legs.

Inflammatory bowel disease

Rheumatoid arthritis and sero-negative arthritides

Haematological: leukaemia, myelofibrosis

Complications:

Perforation
Fistula formation
Haemorrhage
Obstruction

Investigations

Simple	Stool mc&s	
Blood	**Hb**	Anaemia of chronic disease/B12/Fe
	WBC	Inflammation
	CRP, ESR	Inflammation
	Culture	If septic
Imaging	**Ba meal + follow through**	

"Rose thorn lesions" =mucosal ulceration
"String sign" = strictures
Skip lesions
Fistulae

Invasive **Endoscopy**

Visualisation + biopsy
"Cobblestoning"
Skip lesions

Treatment

Conservative

Elemental diet if severe, dietician review
Nutritional support (may require B12/Fe supplements)

Medical

Symptomatic

Anti-diarrhoeals (only in chronic situation)

Anti-inflammatory

Steroids
5-ASAs
Azathioprine
Methotrexate
Infliximab

Interventional

80% of Crohn's patients will require surgery at some point

Resection is conservative, aiming to keep as much bowel as possible to avoid short bowel syndrome (malabsorption due to insufficient bowel).

Indications

Failure of medical therapy

Complications such as perforation, fistulae

Prognosis

Relapses are almost inevitable, with close to 100% recurrence over 20 years.

Ulcerative Colitis

Definition

Chronic recurrent inflammatory disease of large bowel always involving the rectum and spreading in continuity proximally to involve a variable amount of the colon, but not spreading beyond the ileo-caecal valve.

Description

Incidence Common (UC>CD), incidence stable.

Age Any, peaks at 20-40

Sex F>M

Geography Worldwide. ↑ West, Whites, Jews

Aetiology

Risk Factors

Family history (HLA B27)

Non-smokers

Autoimmune disease

Pathophysiology

Macropathology

Inflammed mucosa that bleeds easily
Continuous lesion
Proctitis with variable spread proximally
Only the colon is affected i.e. to ileocaecal valve, although "backwash ileitis" may occur.

Micropathology

Mucosal inflammation (not transmural)
Crypt abscesses
Goblet cell depletion

History and Examination

Exacerbations and remissions of:

Diarrhoea + blood + mucus ± urgency and tenesmus
Abdominal pain
Blood on PR
Clubbing

Systemic features:

Constitutional	Fever, malaise, weight loss
Eyes	Conjunctivitis episcleritis, iritis
Joints	Arthralgia, sacroilitis, ankylosing spondylitis
Skin	Erythema nodosum, pyoderma gangrenosum
Liver	Fatty liver, gallstones, sclerosing cholangitis
	Cholangiocarcinoma
Renal	Stones

Severe attack:

> 6 stool/ day + blood
> 37.5°C pyrexia
> 90 bpm tachycardia
> 30 ESR
< 10g/dL Hb
<30g/dL albumin

Complications:

Toxic megacolon

Dilatation of colon which may perforate, ***surgical emergency***

Malignancy

10% risk colon CA for every 10 years of UC

Haemorrhage and perforation are rare as ulceration is superficial

Investigations

Simple	Urine dipstick	UTI
	Stool mc&s	
Blood	**Hb**	Anaemia (Chronic disease/ Fe deficiency)
	WBC	Inflammation
	CRP, ESR	Inflammation
	LFTs	Albumin (severity) Primary sclerosing cholangitis
	Culture	If septic
Imaging	**AXR**	**Colonic dilatation**
	Ba enema	Lead pipe appearance
Invasive	**Endoscopy**	Visualisation + biopsy

Differential diagnosis of Colitis (= inflammation of the colon)

Inflammatory	Crohn's disease/ Ulcerative colitis
Infective	*Bacterial*: Campylobacter, salmonella, shigella, E. coli V*iral:* Enterovirus, CMV, herpes *Protozoal,* Entamoeba histolytica
Pseudomembranous	C. difficile
Ischaemic	
Radiation	

Treatment

Conservative

NBM if severe acute attack, dietician review, nutritional support

Medical

Mild

Oral/rectal prednisolone

Oral/ rectal 5-ASA e.g. mesalazine

Severe

Hydrocortisone i.v. /enema

5-ASA

Azathioprine

Cyclosporin

Maintenance

5-ASA reduces frequency of flare ups

Interventional

Indications

Failure of medical therapy

Severe exacerbations of colitis/ non-GI manifestations-(rare)

Toxic colon/ acute dilatation

Development of dysplasia/ carcinoma

Proctocolectomy + permanent ileostomy

All diseased/ potentially diseased bowel removed

Colectomy + temporary ileostomy + subsequent ileorectal anastamosis

Rectum remains and may become diseased + ↑ risk CA

Ileal pouch + ileoanal anastamosis

Ileum folded back to create a pouch which can store stool as a makeshift rectum
Continence maintained
May be complicated by "pouchitis"

Prognosis

Mostly good prognosis if only proctitis -10% develop colitis.

Severe disease has 10-20% mortality which is decreased with prompt treatment and surgery.

Crohn's vs. Ulcerative Colitis

Crohn's	Ulcerative Colitis
Macropathology Skip lesions Mouth to anus	**Macropathology** Continuous lesion Anus → proximal spread
Micropathology Full thickness Non-caesiating granuloma	**Micropathology** Partial thickness Crypt abscesses
Presentation Painful diarrhoea Malabsorption	**Presentation** Bloody diarrhoea
Complications Strictures Fistulae Perforation	**Complications** Toxic megacolon
Radiology (Ba studies) Rose thorn ulcers	**Radiology (Ba studies)** Lead piping
Treatment 65% require surgery Limited resections to avoid short bowel syndrome	**Treatment** 85% managed medically Protocolectomy curative (but not without its problems)

COELIAC DISEASE

Definition

Small bowel mucosal sensitivity to gluten in genetically predisposed individuals. Enteropathy improves on a gluten-free diet and relapses on reintroduction.

Description

Incidence	Common 1/1500 in UK (1/300 are positive with serological screening)
Age	Any, peaks at 0-5 and 30-40 years
Sex	F>M
Geography	↑ Incidence Europe, especially Ireland, Whites (important for MCQs!)

Aetiology

Risk factors

Family history: 10-15% of first degree relatives

Associated with atopy and autoimmune disease HLA DR3/DQ2

Pathophysiology

Mucosal damage of proximal small bowel as gluten is broken down to gliadin. A complex, immunological reaction to gliadin peptides results in subtotal villous atrophy. Villus atrophy coupled with crypt hyperplasia results in a smooth mucosal surface which can be seen on histology.

History and Examination

The majority of people are essentially asymptomatic. There is an association with other autoimmune diseases such as Type I Diabetes (5% have coeliac disease) and Grave's.

Symptoms are non-specific.

Fatigue, malaise
Diarrhoea/ steatorrhoea
Abdominal pain
Weight loss

Dermatitis herpetiformis = subempidermal blistering + Coeliac disease

Complications

Anaemia, osteoporosis

Angular stomatitis (mouth ulcers)

Paraesthesia, muscle weakness, polyneuropathy

Malignancy: lymphoma, GI (only if coeliac not treated)

Gross malnutrition (rare)

Investigations

Simple	**Eliminate gluten then challenge** Note: Only *after* biopsy /endomysial antibodies	
Blood	**Hb**	Anaemia (Fe/Folate/B12)
	Endomysial Abs	Sensitive and specific
	Tissue Transgutaminase Abs	Sensitive and specific
Invasive	**Endoscopy** (diagnostic)	Jejunal biopsy

Treatment

Conservative

Gluten free diet, avoid wheat, barley and rye

Medical

Folate/ Fe to treat anaemia

Prognosis

Good if compliant with gluten-free diet where the small bowel mucosa (either in part or completely) returns to normal histology. Increased risk of malignancy only in those untreated for less than 5 years.

INDEX

A

B

C

D

E

F

G

H

I

J

K

L

M

N

O

P

Q

R

S

T

U

V

W

X

Z

About the Authors

This book was written in the authors' FY2 year. They are now ST3 doctors who have been involved in undergraduate medical education for over eight years.

Cristina is a neurology registrar and Honorary Lecturer at Imperial College School of Medicine. She has worked as an examiner as well as on e-Learning via WebCT and designing virtual patients. Andreas is a plastic surgery registrar who has demonstrated anatomy at Brighton and Sussex Medical School and Imperial College.

Thinking medicine was written by young doctors who understand students' needs and examiners' expectations.

Lightning Source UK Ltd.
Milton Keynes UK
UKOW05f2318111013

218959UK00001B/95/P

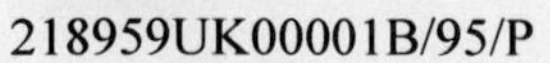